Finite Element Analysis for Undergraduates

Finite Element Analysis for Undergraduates

J. Ed Akin

Professor and Chairman, Department of Mechanical Engineering and Materials Science, Rice University, Houston, Texas, USA

ACADEMIC PRESS

Harcourt Brace Jovanovich, Publishers

London · San Diego · New York · Berkeley
Boston · Sydney · Tokyo · Toronto

ACADEMIC PRESS LIMITED
24–28 Oval Road, London NW1 7DX

United States Edition published by
ACADEMIC PRESS, INC.
San Diego, CA 92101

Second printing (pbk only) 1988
Third printing (pbk only) 1989

British Library Cataloguing in Publication Data

Akin, J.E.
Finite element analysis for undergraduates
1. Finite element method
I. Title
515.3'53 TA347.F5

ISBN 0-12-047655-X
ISBN 0-12-047656-8 (Pbk)

Printed in Great Britain by
St Edmundsbury Press Limited, Bury St Edmunds, Suffolk

Preface

This text is intended to introduce the topic of finite element analysis to undergraduate students. It requires the minimum amount of prerequisite material. These include the first courses in calculus, physics, and mechanics of solids. An introduction to matrix algebra is included. Some FORTRAN programs have been included to illustrate the implementation of matrix algebra.

The material covered progresses from physical intuition in one dimension to mathematical formulations in one and two dimensions. The text was designed for a semester system. Some sections have been marked with an asterisk to denote that they could be omitted in a quarter system.

I would like to thank my family for their support and sacrifices during the time the text was being prepared. The efforts of Ms Linda Anderson are gratefully acknowledged. She prepared the manuscript on the NBI word-processor. The text has been directly photocopied from the output of that system.

Houston
August 1986

J. E. Akin

To my parents:
Mr John Akin and Mrs Ruth Willis

Contents

1. INTRODUCTORY CONCEPTS

1.1 Introduction

The digital computer has become a common analysis tool for engineers. It is often employed in the solution of differential equations that arise in engineering. The two most common techniques utilized to solve these problems are finite difference methods and finite element methods. Both of these methods replace the original differential equations by a set of algebraic equations. A brief comparison of these two approaches will now be outlined. The finite difference method is older and well understood. One usually begins the analysis by approximating the region of interest with a grid of uniformly spaced nodes. At each of these nodes each derivative in the governing differential equation is approximated by an algebraic expression which references the adjacent node points. The system of algebraic equations is obtained by evaluating the above expressions at each node on the grid. Then the system of equations is solved for the value of the dependent variable at each node.

The finite element method is a more recent approach that is now well established. One begins the analysis by approximating the region of interest by subdividing it into a number of non-uniform finite elements that are connected to associated nodes. Within each typical element the change of the dependent variable with location is approximated by an interpolation function. This function is defined relative to the values of the dependent variable at the nodes associated with the element. The original problem is then replaced with some type of equivalent integral statement. This is a significant difference in the two approaches. Next the assumed interpolation functions are substituted into the governing integral form, integrated, and combined with the results from other elements. This yields the governing algebraic equations to be solved for the dependent variable at each node.

When these two approaches are considered in detail a number of relative advantages and disadvantages can be noted. The arbitrary mesh grading that is possible with a finite element formulation allows simultaneous treatment of both detailed representation in certain locations combined with a coarse representation in regions of less interest. This is achieved with a standard finite element formulation, whereas special tricks are often needed for non-uniform finite difference mesh transitions. The finite difference method usually requires special modifications to define the location of points on an arbitrarily shaped boundary. The boundary conditions are usually easily represented in the finite element method of analysis. However, the analysis by finite differences often requires the introduction of fictitious boundary regions in order to satisfy the boundary conditions. The ability to handle inhomogeneous problems are available in a standard finite element formulation. In the finite difference method special interface conditions must be considered where sudden changes in properties occur.

The nodal error associated with a finite difference analysis can be closely estimated. Convergence requirements for the finite element analysis are not as completely established. However, the main sources of error are easily controlled. Thus, the finite element method appears to have a number of important practical advantages for the solution of engineering problems.

The finite element method is an important and practical numerical analysis tool. It has found application in almost all areas of engineering and applied mathematics. The literature on finite element methods is extensive and rapidly increasing. Numerous texts are available which present the theory of various finite element procedures. Most of these are specifically designed to be presented at a graduate level engineering or mathematics course. Thus, most of the available references are not suitable for use in an undergraduate program. By way of comparison this presentation is designed with third or fourth year undergraduate students in mind. Still, new concepts will be introduced as we proceed through the subject.

Since the finite element method usually relies heavily on the computer, a few sample programs are included to illustrate some computational considerations. They often present algorithms written for clarity rather than efficiency. A few texts present a more detailed study of related computation procedures [1], [13]. This chapter will introduce some of the basic concepts and terminology associated with modern finite element methods. A number of the concepts will be expanded in detail in later sections.

1.2 Foundation of Finite Element Analysis

From the mathematical point of view the finite element method is based on integral formulations. By way of comparison the older finite difference methods are usually based on differential formulations. Finite element models of various problems have been formulated from simple physical intuition and from mathematical principles. Historically, the use of physical intuition led to several early practical models. Today there is increased emphasis on the well established mathematical foundations of the procedure. However, since we wish to begin with an elementary presentation we will begin by using physical intuition. The following sections will briefly review the common procedures for establishing finite element models. It is indeed fortunate that all of these techniques use the same **bookkeeping** operations to generate the final assembly of algebraic equations that must be solved for the unknown nodal parameters.

The earliest mathematical formulations for finite element models were based on variational techniques. Variational techniques still are very important in developing elements and in solving practical problems. This is especially true in the areas of structural mechanics and stress analysis. Modern analysis in these areas has come to rely on finite element techniques almost exclusively. **Variational models** usually involve finding the nodal parameters that yield a stationary (maximum or minimum) value of a specific integral relation known as a functional. In most cases it is possible to assign a physical meaning to the integral being extremised. For example, in solid mechanics the integral may represent potential energy, whereas in a fluid mechanics problem it may correspond to the rate of entropy production.

The generation of finite element models by the utilization of **weighted residual** techniques is a relatively recent development. However, these methods are increasingly important in the solution of differential equations and other non-structural applications. The weighted residual method starts with the governing differential equation and avoids the search for a mathematically equivalent variational statement. Generally one assumes an approximate solution, and substitutes this solution into the differential equation. Since the assumption is approximate, this operation defines a residual error term, R, in the differential equation. Although one cannot force the residual term to vanish, it is possible to force a weighted integral of the residual to vanish. That is, the integral over the solution domain, of the product of the residual term and some weighting function, is set equal to zero. Therefore,

$$I = \int RW \, dv = 0.$$

This gives us a direct way to express an approximate solution as an integral for use in finite element solutions.

1.3 General Finite Element Analysis Procedure

In the finite element method, the boundary and interior of the continuum (or more generally the solution domain) are subdivided by points, lines or surfaces into a finite number of discrete sized subregions or **finite elements.** A discrete number of **nodal points** are established with the mesh that divides the region. These nodal points can lie anywhere along, or inside, the subdividing mesh lines, but they're usually located at intersecting mesh lines (or surfaces). Usually, the elements have straight boundaries and thus some geometric approximations will be introduced in the geometric idealization if the actual region of interest has curvilinear boundaries.

The nodal points are assigned identifying integer numbers **(node numbers).** Similarly, each element is assigned an identifying integer number. These **element numbers** begin with unity and extend to a maximum value. As will be discussed later, the assignment of the nodal numbers and element numbers can have a significant effect on the solution time and storage requirements. The analyst assigns a number of (generalized) **degrees of freedom,** (dof), to every node. These are the (unknown) nodal parameters that have been chosen by the analyst to govern the formulation of the problem of interest. Common **nodal parameters** are pressure, velocity components, displacement components, displacement gradients, etc. The nodal parameters do not have to have a physical meaning, although they usually do. A typical node, Fig. 1.1, will usually be associated with more than one element. The domains of influence of a typical node and typical element are also shown in Fig. 1.1. A typical element will have a number of nodal points associated with it located on or within its boundaries. This idealization procedure defines the total number of degrees of freedom associated with a typical node and a typical element. Obviously, the number of degrees of freedom in the system, say **NSF**, is the product of the number of nodes and the number of parameters per node. Similarly, the number of degrees of freedom per element, say **NLF**, is defined by the product of the number of nodes per element and the dof per node. Certain simple bookkeeping operations will be required to relate the element dof and the system dof.

Data must be supplied to define the spatial coordinates of each nodal point. It is common to associate an integer code with each nodal point. The purpose of the code is to indicate which, if any, of the nodal parameters at the node have boundary constraints specified. An important concept is that of **element connectivity**, i.e. the list of global node numbers that are attached to an element. The element connectivity data defines the **topology** of the mesh. Thus for each element, it is necessary to input (in some consistent order) the node numbers that are associated with that particular element. This array of data, say NODES, provides the information necessary to execute certain important bookkeeping steps. A fundamental problem is to establish the element matrices. Generally, they involve substituting interpolation functions into the governing integral form. Historically, the resulting matrices have been called the **element stiffness matrix** and **load vector**, respectively. Although these matrices can sometimes be developed from physical intuition, they are usually formulated by the minimization of a functional or by the method of weighted residuals. These procedures are described in several texts and will be illustrated in later chapters.

Almost all element matrix definitions involve some type of defining properties, or coefficients. A few finite element problems require the definition of properties at the nodal points. For example, in a stress analysis one may wish to define variable thickness elements by specifying the thickness of the material at each node point. The finite element method is very well suited to the solution of non-homogeneous problems; therefore, most finite element programs also require the analyst to assign certain properties to each element.

Once the element equations have been established the contribution of each element is added to form the **system equations.** The system equations resulting from a finite element analysis will usually be symmetric. After the system equations have been assembled, it is necessary to apply the boundary constraints before solving for the unknown nodal parameters. The most common types of nodal parameter boundary **constraints** are

(1) defining explicit values of the unknowns on the boundary, and

(2) defining constraint equations that are linear combinations of the nodal quantities.

Methods for accomplishing these conditions will be presented later. Another type of boundary condition is the type that involves a flux or traction on the boundary of one or more elements. These element boundary constraints contribute additional terms to the element square and/or column matrices for the element(s) on which the constraints were placed. Thus, although these (Neumann type) conditions do enter into the system equations, their presence may not be obvious at the system level.

After all the above conditions are satisfied, the system equations are solved by means of procedures which account for the sparse, symmetric nature of the problem. This can greatly reduce (by 90% or more) the number of calculations which would normally be required to solve the equations. After the equations have been solved for the unknown nodal parameters, it is usually necessary to output those parameters. In some cases the problem would be considered completed at this point, but in others it is necessary to use the calculated values of the nodal parameters to calculate other quantities of interest. For example, in stress analysis one uses the calculated nodal displacements to solve for the stresses and strains. Of course, these secondary quantities must also be output in some manner. Techniques of this type are usually called **post-processing.**

One related task that also often arises with the finite element technique is the problem of data generation. One disadvantage of finite element procedures is that they can require large amounts of data. For regions with complex geometries, the requirements of closely approximating the boundaries together with the desirability of a fine mesh make it desirable to use a large number of nodal points and elements. The specification of the locations of such a large number of nodal points and elements is a time consuming job in which there is a high probability of human error. Data preparation and evaluation can require as much as thirty to forty percent of the total cost and time involved in the solution of large practical problems. To minimize the data preparation time and the probability of error several schemes for the automatic generation of much of the required data have been developed. These operations are usually called **pre-processing.**

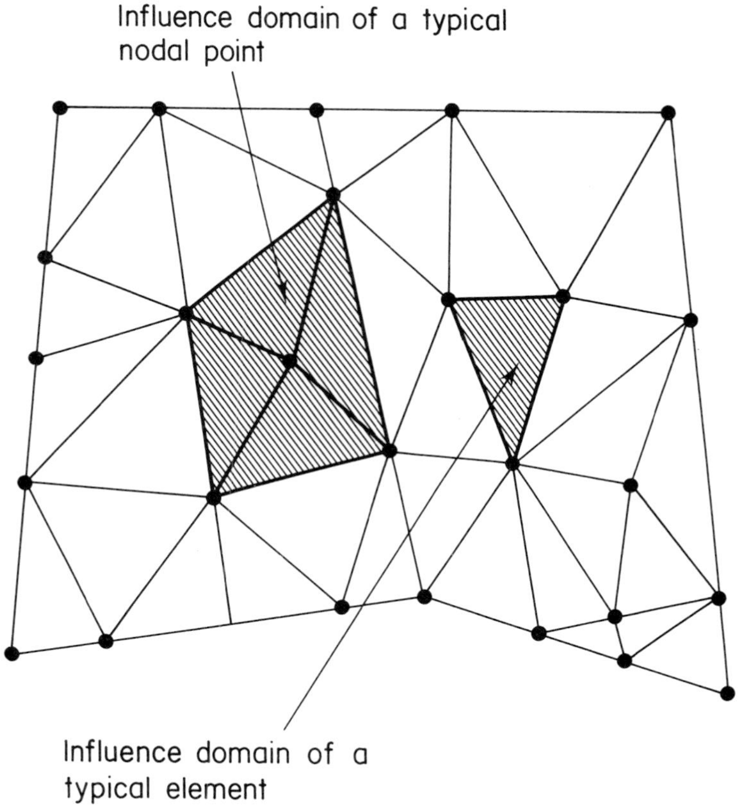

Fig. 1.1 Influence Domains

2. MATRIX NOTATION AND OPERATIONS

2.1 Introduction

Finite element analysis procedures are most commonly described using matrix notation. These procedures ultimately lead to the solution of a large set of simultaneous equations. Thus for the sake of completeness and consistency with later sections, the basic forms of matrix notation and matrix operations will be described here.

2.2 Matrices

A **matrix** is defined as a rectangular array of quantities arranged in rows and columns. The array is enclosed in brackets, and thus if there are **m** rows and **n** columns, the matrix can be represented by

$$A = \begin{bmatrix} a_{11} & a_{12} & a_{13} & \cdots & a_{1j} & \cdots & a_{1n} \\ a_{21} & a_{22} & a_{23} & \cdots & a_{2j} & \cdots & a_{2n} \\ a_{31} & a_{32} & a_{33} & \cdots & a_{3j} & \cdots & a_{3n} \\ \cdots & \cdots & \cdots & \cdots & \cdots & \cdots & \\ a_{i1} & a_{i2} & a_{i3} & \cdots & a_{ij} & \cdots & a_{in} \\ \cdots & \cdots & \cdots & \cdots & \cdots & \cdots & \\ a_{m1} & a_{m2} & a_{m3} & \cdots & a_{mj} & \cdots & a_{mn} \end{bmatrix} = [A] \qquad (2.1)$$

where the typical element a_{ij} has two subscripts, of which the first denotes the row (i-th) and the second denotes the column (j-th) which the element occupies in the matrix. A matrix with **m** rows and **n** columns is defined as a matrix of order **m** x

n, or simply an **m** x **n** matrix. The number of rows is always specified first. In Eq. (2.1) the symbol **A** stands for the matrix of **m** rows and **n** columns and it is usually printed in **boldface** type. If **m** = **n** = 1 then the matrix degenerates to a scalar. If **m** = 1, the matrix **A** reduces to the single row

$$\mathbf{A} = [a_{11}\ a_{12}\ a_{13}\ \cdots\ a_{1j}\ \cdots\ a_{1n}] = (A) \tag{2.2}$$

which is called a **row matrix.** Similarly, if **n** = 1, the matrix A reduces to the single column

$$\mathbf{A} = \begin{Bmatrix} a_{11} \\ a_{21} \\ \cdot \\ \cdot \\ \cdot \\ \cdot \\ a_{m1} \end{Bmatrix} = \{A\} \tag{2.3}$$

which is called a **column matrix.** When all the elements of matrix are equal to zero, the matrix is called **null** or **zero** and is indicated by 0. A null matrix serves the same function as zero does in ordinary algebra. If **m** = **n**, the matrix **A** is the square array

$$\mathbf{A} = \begin{bmatrix} a_{11} & a_{12} & \cdots & a_{1n} \\ & & \cdots\cdots & \cdots \\ a_{n1} & a_{n2} & \cdots & a_{nn} \end{bmatrix} \tag{2.4}$$

which is called a square matrix. A system of linear equations may be written in the general form

$$\begin{aligned} a_{11}x_1 + a_{12}x_2 + a_{13}x_3 + \cdots + a_{1n}x_n &= c_1 \\ a_{21}x_1 + a_{22}x_2 + a_{23}x_3 + \cdots + a_{2n}x_n &= c_2 \\ a_{31}x_1 + a_{32}x_2 + a_{33}x_3 + \cdots + a_{3n}x_n &= c_3 \\ &\vdots \\ a_{n1}x_1 + a_{n2}x_2 + a_{n3}x_3 + \cdots + a_{nn}x_n &= c_n \end{aligned}$$

where the a_{ij}'s and c_i's represent known coefficients and the x_i's are unknowns. These equations may also be represented in the more compact form given by

$$\mathbf{AX} = \mathbf{C} \tag{2.5}$$

where **A**, **X**, and C represent matrices. In expanded form this is

$$\begin{bmatrix} a_{11} & a_{12} & a_{13} & \cdots & a_{1n} \\ a_{21} & a_{22} & a_{23} & \cdots & a_{2n} \\ a_{31} & a_{32} & a_{33} & \cdots & a_{3n} \\ \cdot & & & & \cdot \\ \cdot & & & & \cdot \\ \cdot & & & & \cdot \\ a_{n1} & a_{n2} & a_{n3} & \cdots & a_{nn} \end{bmatrix} \begin{Bmatrix} x_1 \\ x_2 \\ x_3 \\ \cdot \\ \cdot \\ \cdot \\ x_n \end{Bmatrix} = \begin{Bmatrix} c_1 \\ c_2 \\ c_3 \\ \cdot \\ \\ \\ c_n \end{Bmatrix}. \tag{2.6}$$

Before considering some of the matrix algebra implied by the above equation, a few other matrix types will be defined for future reference.

A **diagonal matrix** is a square matrix which has zero elements outside the principal diagonal. It follows therefore that for a diagonal matrix $a_{ij} = 0$ when $i \neq j$ and not all a_{ii} are zero. A typical diagonal matrix may be represented by

$$\mathbf{A} = \begin{bmatrix} a_{11} & 0 & \cdots & 0 \\ 0 & a_{22} & \cdots & 0 \\ \cdots & \cdots & \cdots & \cdots \\ 0 & 0 & \cdots & a_{nn} \end{bmatrix} \tag{2.7}$$

or simply, in order to save space, as

$$\mathbf{A} = \lceil a_{11} \quad a_{22} \quad \cdots \quad a_{nn} \rfloor = \lceil A \rfloor .$$

A **unit matrix** is a square matrix whose elements are equal to 0 except those located on its main diagonal, which are equal to 1. That is: $a_{ij} = 1$ if $i = j$ and $a_{ij} = 0$ if $i \neq j$. The unit matrix will be given the symbol **I** throughout this book. An example of a 3 x 3 unit matrix is:

$$\mathbf{I} = \begin{bmatrix} 1 & 0 & 0 \\ 0 & 1 & 0 \\ 0 & 0 & 1 \end{bmatrix} .$$

A **symmetric matrix** is a square matrix whose elements $a_{ij} = a_{ji}$ for $i \neq j$. For example:

$$\mathbf{S} = \begin{bmatrix} 12 & 2 & -1 \\ 2 & 33 & 0 \\ -1 & 0 & 15 \end{bmatrix} .$$

An **anti-symmetric matrix** (skew symmetric) is a square matrix whose elements $a_{ij} = -a_{ji}$ for $i \neq j$. , and $a_{ii} = 0$. For example:

$$\mathbf{A} = \begin{bmatrix} 0 & 2 & -1 \\ -2 & 0 & 10 \\ 1 & -10 & 0 \end{bmatrix} .$$

The **transpose** $\mathbf{A}^T$, of a matrix **A** is obtained by interchanging the rows and columns. Thus the transpose of an **m x n** matrix is an **n x m** matrix. For example:

$$\text{if} \quad \mathbf{A} = \begin{bmatrix} 2 & 1 \\ 3 & 5 \\ 0 & 1 \end{bmatrix} \quad \text{then} \quad \mathbf{A}^T = \begin{bmatrix} 2 & 3 & 0 \\ 1 & 5 & 1 \end{bmatrix} .$$

Column matrices are often written transposed so as to save space.

If all the elements on one side of the diagonal of a square matrix are zero, the matrix is called a **triangular matrix.** There are two types of triangular matrices; (1) an upper triangular **U** whose elements below the diagonal are all zero, and (2) a lower triangular **L**, whose elements above the diagonal are all zero. An example of a lower triangular matrix follows:

$$\mathbf{L} = \begin{bmatrix} 10 & 0 & 0 \\ 1 & 3 & 0 \\ 5 & 1 & 2 \end{bmatrix} .$$

A matrix may be divided into smaller arrays by horizontal and vertical lines. Such a matrix is then referred to as a **partitioned matrix,** and the smaller arrays are called **submatrices.** For example,

$$\mathbf{A} = \begin{bmatrix} a_{11} & a_{12} & \cdot & a_{13} \\ a_{21} & a_{22} & \cdot & a_{23} \\ \cdots & \cdots & \cdots & \cdots \\ a_{31} & a_{32} & \cdot & a_{33} \end{bmatrix} = \begin{bmatrix} \mathbf{A}_{11} & \mathbf{A}_{12} \\ \mathbf{A}_{21} & \mathbf{A}_{22} \end{bmatrix}$$

$$= \begin{bmatrix} 2 & 1 & . & 3 \\ 10 & 5 & . & 0 \\ \cdots & \cdots & \cdots & \cdots \\ 4 & 6 & . & 10 \end{bmatrix} \qquad (2.8)$$

where

$$\begin{aligned} \mathbf{A}_{11} &= \begin{bmatrix} a_{11} & a_{12} \\ a_{21} & a_{22} \end{bmatrix} = \begin{bmatrix} 2 & 1 \\ 10 & 5 \end{bmatrix} \\ \mathbf{A}_{12}^{T} &= \begin{bmatrix} a_{13} & a_{23} \end{bmatrix} = \begin{bmatrix} 3 & 0 \end{bmatrix} \\ \mathbf{A}_{21} &= \begin{bmatrix} a_{31} & a_{32} \end{bmatrix} = \begin{bmatrix} 4 & 6 \end{bmatrix} \\ \mathbf{A}_{22} &= \begin{bmatrix} a_{33} \end{bmatrix} = \begin{bmatrix} 10 \end{bmatrix} . \end{aligned} \qquad (2.9)$$

It should be noted that the elements of a partitioned matrix must be so ordered that they are compatible with the whole matrix A and with each other. That is, $\mathbf{A}_{11}$ and $\mathbf{A}_{12}$ must have an equal number of rows. Likewise $\mathbf{A}_{21}$ and $\mathbf{A}_{22}$ must have an equal number of rows. Matrices $\mathbf{A}_{11}$ and $\mathbf{A}_{21}$ must have an equal number of columns. Likewise for $\mathbf{A}_{12}$ and $\mathbf{A}_{22}$. It should be noted that $\mathbf{A}_{22}$ is a matrix even if it consists of only one element. Provided the general rules for matrix algebra are observed, the submatrices can be treated as if they were ordinary matrix elements.

2.3 **Matrix Algebra**

Addition of two matrices of the same order is accomplished by adding corresponding elements of each matrix. The matrix addition C = **A** + B, where **A, B** and **C** are matrices of the same order **m** x **n** can be indicated by the equation

$$c_{ij} = a_{ij} + b_{ij} \qquad (2.10)$$

where c_{ij}, a_{ij}, and b_{ij} are typical elements of the **C, A,** and **B** matrices, respectively. A typical matrix addition is:

$$\begin{bmatrix} 3 & 0 & 1 \\ 2 & -1 & 2 \\ 1 & 1 & 1 \end{bmatrix} + \begin{bmatrix} -1 & 1 & -1 \\ 2 & 5 & 6 \\ -3 & 4 & 9 \end{bmatrix} = \begin{bmatrix} 2 & 1 & 0 \\ 4 & 4 & 8 \\ -2 & 5 & 10 \end{bmatrix} .$$

The finite element procedures that will be developed will depend heavily on the use of the computer to manipulate matrices. Thus, now is a good time to begin thinking of equations like Eq. (2.10) from the point of view of a scientific language like FORTRAN. Therefore, we should note that any matrix would be a subscripted variable and should be listed in a DIMENSION statement. Similarly Eq. (2.10) implies a range over all values of the subscripts i and j. This would be accomplished with DO loops. To force one to be more conscious of this goal of utilizing the computer some standard shorthand mathematical notation will be used. For example, $\forall$ means "for every". Thus the notation $\forall\ i:\ 1 \leq i \leq m$ would be read as "for every i ranging from 1 to m". Therefore, Eq. (2.10) could be rewritten as:

$$c_{ij} = a_{ij} + b_{ij} \qquad \begin{array}{l} \forall\ i:\ 1 \leq i \leq m \\ \forall\ j:\ 1 \leq j \leq n. \end{array}$$

This more clearly implies the required DO loops in a matrix addition subroutine such as Fig. 2.1.

```
      SUBROUTINE  MATADD (M,N,A,B,C)
C-->  MATRIX ADDITION, C = A + B
      DIMENSION A(M,N), B(M,N), C(M,N)
      DO  5 J = 1,N
      DO 10 I = 1,M
10    C(I,J) = A(I,J) + B(I,J)
5     CONTINUE
      RETURN
      END
```

Fig. 2.1 A Typical Matrix Addition Routine

One may want to note the economy considerations (in Fig. 2.1) that columns add faster than rows and that subscripts should range on different line numbers to reduce termination checks. Matrix subtraction, **C** = **A** - **B**, is performed in a similar manner:

$$c_{ij} = a_{ij} - b_{ij} \qquad \begin{array}{l} \forall\ i:\ 1 \leq i \leq m \\ \forall\ j:\ 1 \leq j \leq n. \end{array}$$

Matrix addition and subtraction are associative and commutative. That is,

$$\mathbf{A} + (\mathbf{B} + \mathbf{C}) = (\mathbf{A} + \mathbf{B}) + \mathbf{C}$$

and

$$\mathbf{A} + \mathbf{B} + \mathbf{C} = \mathbf{C} + \mathbf{B} + \mathbf{A}.$$

This means that the order in which addition and subtraction is carried out has no effect on the result. Multiplication of the matrix **A** by a scalar c is defined as the multiplication of every element of the matrix by the scalar c. Thus the elements of the product $\mathbf{B} = c\mathbf{A}$ are given by $b_{ij} = ca_{ij}$.

Two matrices **A** and **B** can be multiplied together in order **AB** only when the number of columns in **A** is equal to the number of rows in **B**. When this condition is fulfilled, the matrices **A** and **B** are said to be conformable for multiplication. Otherwise, matrix multiplication is not defined. The product of two conformable matrices **A** and **B** of order **m** x **p** and **p** x **n**, respectively, is defined as a matrix $\mathbf{C} = \mathbf{AB}$ of order **m** x **n** calculated from

$$c_{ij} = \sum_{k=1}^{p} a_{ik} b_{kj} \qquad \begin{matrix} \forall\ i:\ 1 \leq i \leq m \\ \forall\ j:\ 1 \leq j \leq n \end{matrix} . \tag{2.11}$$

Notice that the appearance of the summation sign, Σ, implies the existence of another DO loop. When evaluating $\mathbf{AB} = \mathbf{C}$ by hand, it is helpful to arrange **A** and **B** matrices as shown in Eq. (2.12). The element c_{ik} which results from multiplying elements of row **i** of the **A** matrix by elements of column **k** of the B matrix will be located by projecting lines through row i of **A** and column k of **B**:

$$\mathbf{B} = \begin{bmatrix} b_{11} & b_{12} & \cdots & b_{1k} & \cdots \\ b_{21} & b_{22} & \cdots & b_{2k} & \cdots \\ \cdot & \cdot & & \cdot & \\ \cdot & \cdot & & b_{nk} & \cdots \end{bmatrix}$$

$$\mathbf{A} = \begin{bmatrix} a_{11} & a_{12} & a_{13} & \cdots & a_{1n} \\ a_{21} & a_{22} & a_{23} & \cdots & a_{2n} \\ \cdot & & & & \\ \cdot & & & & \\ a_{i1} & a_{i2} & a_{i3} & \cdots & a_{in} \\ \cdot & & & & \\ \cdot & & & & \end{bmatrix} \qquad \begin{bmatrix} & & & \cdot & \\ & & & \cdot & \\ & & & \cdot & \\ & & & \cdot & \\ \cdot & \cdot & \cdot & c_{ik} & \\ & & & & \\ & & & & \end{bmatrix}$$

and the element c_{ik} is given in expanded form as:

$$c_{ik} = a_{i1}b_{1k} + a_{i2}b_{2k} + a_{i3}b_{3k} + \dots + a_{ij}b_{jk} + \dots + a_{in}b_{nk}.$$

Matrix multiplication is associative and distributive. For example:

$$(\mathbf{AB})\mathbf{C} = \mathbf{A}(\mathbf{BC})$$

$$\mathbf{A}(\mathbf{B} + \mathbf{C}) = \mathbf{AB} + \mathbf{AC}$$

However, matrix multiplication is not commutative. In general:

$$\mathbf{AB} \neq \mathbf{BA} .$$

Consequently, the order in which matrix multiplication is specified should not be changed. When two matrices A and B are multiplied, the product **AB** is referred to either as B premultiplied by A or as **A** postmultiplied by B. There may be cases when **AB** = **BA**, and the matrices A and B are then said to be **commutable.** For example, the unit matrix I commutes with any square matrix of the same order, that is, **AI** = **IA** = **A.**

The process of matrix multiplication can also be extended to partitioned matrices, provided the individual products of submatrices are conformable for multiplication. For example, the multiplication

$$\mathbf{AB} = \begin{bmatrix} \mathbf{A}_{11} & \mathbf{A}_{12} \\ \mathbf{A}_{21} & \mathbf{A}_{22} \end{bmatrix} \begin{bmatrix} \mathbf{B}_{11} & \mathbf{B}_{12} \\ \mathbf{B}_{21} & \mathbf{B}_{22} \end{bmatrix}$$

$$= \begin{bmatrix} \mathbf{A}_{11}\mathbf{B}_{11} + \mathbf{A}_{12}\mathbf{B}_{21} & \mathbf{A}_{11}\mathbf{B}_{12} + \mathbf{A}_{12}\mathbf{B}_{22} \\ \mathbf{A}_{21}\mathbf{B}_{11} + \mathbf{A}_{22}\mathbf{B}_{21} & \mathbf{A}_{21}\mathbf{B}_{12} + \mathbf{A}_{22}\mathbf{B}_{22} \end{bmatrix}$$

is possible provided the products $\mathbf{A}_{11}\mathbf{B}_{11}$, $\mathbf{A}_{12}\mathbf{B}_{21}$, etc., are conformable. For this condition to be fulfilled, it is only necessary for the vertical partitions in **A** to include a number of columns equal to the number of rows in the corresponding horizontal partitions in B.

Often in finite element studies one encounters the transpose of a product of matrices. The following relationship can be shown to be true:

$$(\mathbf{AB})^T = \mathbf{B}^T\mathbf{A}^T \tag{2.13}$$

or more generally

$$(\mathbf{ABC...YZ})^T = \mathbf{Z}^T\mathbf{Y}^T \; ... \; \mathbf{C}^T\mathbf{B}^T\mathbf{A}^T.$$

A typical algorithm for forming the matrix product in Eq. (2.11) is shown in Fig. 2.2.

```
      SUBROUTINE MATMLT (L,M,N,A,B,C)
C-->  MATRIX PRODUCT, C(L,N)=A(L,M)*B(M,N)
      DIMENSION A(L,M), B(M,N), C(L,N)
      DO 30 I = 1,L
      DO 20 J = 1,N
      SUM = 0.0
      DO 10 K = 1,M
10    SUM = SUM + A(I,K)*B(K,J)
20    C(I,J) = SUM
30    CONTINUE
      RETURN
      END
```

Fig. 2.2 A Typical Matrix Multiplication Routine

As an example of matrix multiplication let $\mathbf{B}^T = [3 \quad 1 \quad 2]$ and let

$$\mathbf{A} = \begin{bmatrix} 2 & 1 & 0 \\ 1 & 0 & 1 \end{bmatrix}$$

then their product $\mathbf{C} = \mathbf{AB}$ is:

$$\begin{matrix} & & \begin{Bmatrix} 3 \\ 1 \\ 2 \end{Bmatrix} & & \\ & & \vdots & & \\ \begin{bmatrix} 2 & 1 & 0 \\ 1 & 0 & 1 \end{bmatrix} & \cdots & \begin{Bmatrix} 7 \\ 5 \end{Bmatrix} = \mathbf{C} = & \begin{Bmatrix} 2*3 + 1*1 + 0*2 \\ 1*3 + 0*1 + 1*2 \end{Bmatrix}. \end{matrix}$$

Every (non-singular) square matrix **A** has an **inverse** which will be indicated by $\mathbf{A}^{-1}$ such that by definition the product $\mathbf{AA}^{-1}$ is a unit matrix

$$\mathbf{AA}^{-1} = \mathbf{I}.$$

The reverse is also true

$$\mathbf{A}^{-1}\mathbf{A} = \mathbf{I}.$$

Inverse matrices are very useful in the solution of simultaneous equations

$$\mathbf{AX} = \mathbf{C}$$

such as Eq. (2.6) where **A** and **C** are known and **X** is unknown. If the inverse of **A** is known, it is possible to solve for the unknowns of the **X** matrix by premultiplying both sides by the inverse

$$\mathbf{A}^{-1}\mathbf{AX} = \mathbf{A}^{-1}\mathbf{C}$$

from which

$$\mathbf{X} = \mathbf{A}^{-1}\mathbf{C} \ . \tag{2.14}$$

Various methods can be used to determine the inverse of a given matrix. For very large systems of equations it is probably more practical to avoid the calculation of the inverse and solve the quations by a procedure called **factorization** [1]. Various procedures for computing an inverse matrix can be found in texts on numerical analysis. The inverse of 2x2 or 3x3 matrices can easily be written in closed form by using **Cramer's rule.** The result for a 3x3 matrix is illustrated in Fig. 2.3.

For future use a general matrix inversion program, INVERT, is included without proof in Fig. 2.4. Note that it requires two extra temporary work arrays, and it replaces the given matrix with its inverse. Later we will find that our major interest will be in systems of equations where the square coefficient matrix is symmetric. This special case allows some simplifications and symmetric matrices can be inverted by using a more efficient algorithm such as SYMINV illustrated in Fig. 2.5.

```
      SUBROUTINE I3BY3 (A,AINV,DET)
C--> FIND INVERSE AND DETERMINATE OF MATRIX A(3,3)
      DIMENSION A(3,3), AINV(3,3)
      AINV(1,1) =   A(2,2)*A(3,3) - A(3,2)*A(2,3)
      AINV(2,1) = - A(2,1)*A(3,3) + A(3,1)*A(2,3)
      AINV(3,1) =   A(2,1)*A(3,2) - A(3,1)*A(2,2)
      AINV(1,2) = - A(1,2)*A(3,3) + A(3,2)*A(1,3)
      AINV(2,2) =   A(1,1)*A(3,3) - A(3,1)*A(1,2)
      AINV(3,2) = - A(1,1)*A(3,2) + A(3,1)*A(1,2)
      AINV(1,3) =   A(1,2)*A(2,3) - A(2,2)*A(1,3)
      AINV(2,3) = - A(1,1)*A(2,3) + A(2,1)*A(1,3)
      AINV(3,3) =   A(1,1)*A(2,2) - A(2,1)*A(1,2)
      DET =  A(1,1)*AINV(1,1) + A(1,2)*AINV(2,1)
     1   +  A(1,3)*AINV(3,1)
      DO 20  J = 1,3
      DO 10  I = 1,3
10    AINV(I,J) = AINV(I,J)/DET
20    CONTINUE
      RETURN
      END
```

Fig. 2.3 Inversion by Cramer's Rule

```
      SUBROUTINE  INVERT  (N,A,B,C)
C--   INVERSION OF NONSYMMETRIC MATRIX A
      DIMENSION A(N,N), B(N), C(N)
C     N = SIZE OF GIVEN MATRIX, N>1
C     B AND C ARE WORKING SPACE VECTORS
C     A IS REPLACED BY ITS INVERSE
      NN=N - 1
      A(1,1) = 1./A(1,1)
      DO 11 M = 1,NN
      K = M + 1
  1   DO 3 I = 1,M
      SUM = 0.0
      DO 2 J = 1,M
  2   SUM = SUM + A(I,J)*A(J,K)
  3   B(I) = SUM
      D = 0.0
      DO 4 I = 1,M
  4   D = D + A(K,I)*B(I)
      D = - D + A(K,K)
      A(K,K) = 1./D
      DO 5 I = 1,M
  5   A(I,K) = - B(I)*A(K,K)
      DO 7 J = 1,M
      SUM = 0.0
      DO 6 I = 1,M
  6   SUM = SUM + A(K,I)*A(I,J)
  7   C(J) = SUM
      DO 8 J = 1,M
  8   A(K,J) = - C(J)*A(K,K)
      DO 10 I = 1,M
      DO 9 J = 1,M
  9   A(I,J) = A(I,J) - B(I)*A(K,J)
 10   CONTINUE
 11   CONTINUE
      RETURN
      END
```

Fig. 2.4 A General Inversion Algorithm

```
      SUBROUTINE SYMINV (A,N)
C--   INVERT SYMMETRIC MATRIX A(N,N)
      DIMENSION A(N,N)
      DO 40 K = 1,N
      D = A(K,K)
      DO 10 J = 1,N
10    A(K,J) = - A(K,J)/D
      DO 30 I = 1,N
      IF (I.EQ.K)  GO TO 30
      DO 20 J = 1,N
      IF (J.EQ.K)  GO TO 20
      A(I,J) = A(I,J) + A(I,K)*A(K,J)
20    CONTINUE
30    A(I,K) = A(I,K)/D
40    A(K,K) = 1.0/D
      RETURN
      END
```

Fig. 2.5 Symmetric Inversion Procedure

2.4 Determinant of a Matrix

Every square matrix, say **A**, has a single scalar quantity associated with it. That scalar is called the determinant, $|\mathbf{A}|$, of the matrix. The determinant is important in solving equations and inverting matrices since it can be shown that the inverse $\mathbf{A}^{-1}$ will not exist if $|\mathbf{A}| = 0$. In such a case the matrix **A** or the set of equations in Eq. (2.5) is said to be **singular.**

This implies that inversion algorithms, like I3BY3 or INVERT, must fail if $|\mathbf{A}| = 0$. Other algorithms are specialized for the sake of economy to work only when $|\mathbf{A}| > 0$. Fortunately, that special case occurs very often in finite element analysis. The beginner should be warned that some algorithms still return results when the system is singular. Usually these answers will be obvious garbage. This occurs, for example, when the problem is physically impossible because of incorrect boundary conditions. For future reference a few facts about determinant will be noted:

1. If two rows or columns are equal then the determinant is zero.
2. Interchanging two rows, or two columns, changes the sign of the determinant.

3. The determinant is unchanged if any row, or column, is modified by adding to it a linear combination of any of the other rows, or columns.
4. A singular square matrix may have nonsingular square partitions.

The last two items will be become significant when we consider how to apply boundary conditions and how to solve a system of equations.

2.5 Matrix Calculus

At times we will find it necessary to differentiate or integrate matrices. These operations are simply carried out on each and every element of the matrix. Let the elements a_{ij} of **A** be function of a parameter t. Then

$$\mathbf{B} = \frac{d\mathbf{A}}{dt}$$

implies

$$b_{ij} = \frac{da_{ij}}{dt} \qquad \begin{matrix} \forall\ i: 1 \leq i \leq m \\ \forall\ j: 1 \leq j \leq n \end{matrix} \tag{2.15}$$

just as

$$\mathbf{C} = \int \mathbf{A}dt$$

implies

$$c_{ij} = \int a_{ij}dt \qquad \begin{matrix} \forall\ i: 1 \leq i \leq m \\ \forall\ j: 1 \leq j \leq n \end{matrix} \quad . \tag{2.16}$$

Clearly, if **A** is a matrix product the matrix operations must be completed before the calculus operations begin.

Often one encounters a scalar matrix, U, defined by a symmetric square **n x n** matrix, **A**, a column vector B, and a column vector of **n** parameters **X**. The usual form is

$$U = {}^1\!/_2\, X^T \mathbf{A}\mathbf{X} + X^T\mathbf{B} + \mathbf{C}. \tag{2.17}$$

When dealing with functional relations the concept of rate of change is often very important. If we have a function, f, of a single independent variable, say x, then we call the rate of change the derivative with respect to x. That quantity is written as df/dx. This concept is easily generalized. Suppose we have a function z of two variables x and y, z = f(x,y). Here we may define two distinct rates of change. One is a rate of change with respect to y, with x being held constant. Thus we can define two **partial derivatives.** When x is allowed to vary the derivative is called the partial derivative with respect to x. This is denoted by $\partial z/\partial x$. By analogy with the usual definition in calculus this can be written as

$$f_x = \frac{\partial f}{\partial x} = \lim_{\Delta x \to 0} \frac{f(x + \Delta x, y) - f(x,y)}{\Delta x} .$$

A similar definition describes the partial derivative with respect to y, Mf/My. The total derivative is the sum of the partial derivatives, e.g.,

$$df = \frac{\partial f}{\partial x} dx + \frac{\partial f}{\partial y} dy.$$

The previous definition can be extended to include a function of any number of independent variables. If we take the derivative of the scalar U with respect to each $X_i \; \forall \; i: 1 \leq i \leq n$ the result is a column vector

$$\frac{\partial U}{\partial \mathbf{X}} = \mathbf{AX} + \mathbf{B} . \tag{2.18}$$

This is verified by expanding Eq. (2.17), differentiating with respect to every X_i in **X** and rewriting the result as a matrix product. An outline of these steps follows:

$$U = \tfrac{1}{2} \sum_{i=1}^{n} \sum_{j=1}^{n} A_{ij} X_i X_j + \sum_{i=1}^{n} X_i B_i + C$$

or

$$\begin{aligned} U = \tfrac{1}{2}\,(&A_{11}X_1^2 + A_{12}X_1X_2 + A_{21}X_2X_1 + \ldots \\ &+ A_{13}X_1X_3 + A_{31}X_3X_1 + A_{33}X_3^2 + \ldots) \\ &+ (X_1B_1 + X_2B_2 + \ldots + X_nB_n) + C. \end{aligned}$$

so that

$$\begin{aligned} \frac{\partial U}{\partial X_1} = \tfrac{1}{2}\,(&2A_{11}X_1 + (A_{12} + A_{21})X_2 + \ldots \\ &+ (A_{13} + A_{31})X_3 + \ldots) + (B_1) + 0 \end{aligned}$$

which from symmetry of **A** reduces to

$$\frac{\partial U}{\partial X_1} = \frac{2}{2}\,(A_{11}X_1 + A_{12}X_2 + A_{13}X_3 + \ldots) + B_1.$$

Similarly for any typical X_i:

$$\frac{\partial U}{\partial X_i} = \sum_{j=1}^{n} A_{ij}X_j + B_i \qquad \forall\ i:\ -1 \leqslant i \leqslant n.$$

Combining all these rows into a set of equations gives the matrix form

$$\frac{\partial U}{\partial \mathbf{X}} = \{\frac{\partial U}{\partial X}\} = \mathbf{AX} + \mathbf{B} \tag{2.19}$$

as cited above.

2.6 Storage of Matrices in FORTRAN

When utilizing the computer to assist us with the manipulation of matrices (or other subscripted variables) it is sometimes necessary to know how an array is actually stored and retrieved. The procedure used to store the elements of an array is to sequentially store all the elements of the first column, then all of the second, etc. Another way of saying this is that the first (left most) subscript ranges over all its values before the second is incremented. After the second subscript has been incremented then the first again ranges over all its values. The main reason to note this concept is that it requires one to use care when extracting submatrices in partitioned arrays. However, the above knowledge can be used to execute some operations more efficiently. For example, the matrix addition procedure could be written as

$$c_k = a_k + b_k \qquad \forall\ k:\ 1 \leqslant k \leqslant m*n.$$

Of course, such a subroutine would have to DIMENSION the arrays with a single subscript.

2.7 Exercises

1. Show that $\begin{bmatrix} 1 & 2 & -1 & 0 \\ 4 & 0 & 2 & 1 \\ 2 & -5 & 1 & 2 \end{bmatrix} - \begin{bmatrix} 3 & -4 & 1 & 2 \\ 2 & -2 & 3 & -1 \\ 2 & -2 & 3 & -1 \end{bmatrix} =$

$$\begin{bmatrix} -2 & 6 & -2 & -2 \\ 2 & 2 & -1 & 2 \\ 0 & -3 & -2 & 3 \end{bmatrix}.$$

2. Verify the products

(a) $[4 \quad 5 \quad 6]\begin{Bmatrix} 2 \\ 3 \\ -1 \end{Bmatrix} = 17$

(b) $\begin{bmatrix} 2 & -1 & 1 \\ 0 & 1 & 2 \\ 1 & 0 & 1 \end{bmatrix}\begin{bmatrix} 2 & -1 & 1 \\ 0 & 1 & 2 \\ 1 & 0 & 1 \end{bmatrix} = \begin{bmatrix} 5 & -3 & 1 \\ 2 & 1 & 4 \\ 3 & -1 & 2 \end{bmatrix}$

(c) $[1 \quad 2 \quad 3]\begin{bmatrix} 4 & -6 & 9 & 6 \\ 0 & -7 & 10 & 7 \\ 5 & 8 & -11 & -8 \end{bmatrix} = [19 \quad 4 \quad -4 \quad -4]$.

3. Use subroutine SYMINV to show that the inverse of

$$\begin{bmatrix} 4 & 3 & 2 & 1 \\ 3 & 4 & 3 & 2 \\ 2 & 3 & 4 & 3 \\ 1 & 2 & 3 & 4 \end{bmatrix} \text{ is } \begin{bmatrix} 0.6 & -0.5 & 0 & 0.1 \\ -0.5 & 1 & -0.5 & 0 \\ 0 & -0.5 & 1 & -0.5 \\ 0.1 & 0 & -0.5 & 0.6 \end{bmatrix}.$$

4. Use subroutine INVERT to solve the above problem.
5. Verify that $\begin{bmatrix} 1 & 2 & 3 \\ 2 & 5 & 7 \\ -2 & -4 & -5 \end{bmatrix}$ is the inverse of $\begin{bmatrix} 3 & -2 & -1 \\ -4 & 1 & -1 \\ 2 & 0 & 1 \end{bmatrix}$.

   by showing their product is the identity matrix.
6. Use subroutine I3BY3 to check the above problem.
7. Use subroutine MATMLT to verify problem 2.C.
8. Write a program, MATSUB, for matrix subtraction and verify problem 1.
9. Write a program, MSMULT, to multiply a matrix by a scalar.
10. Write a program, MTMULT, to calculate $\mathbf{C} = \mathbf{A}^T\mathbf{B}$ when **A** and **B** are given.

11. Write a subroutine to print the rows and columns of a matrix of real (floating point) numbers, say RPRINT.
12. Write a subroutine to zero a matrix, say ZERO.
13. Solve the system of equations **AX** = **B** where

$$\begin{bmatrix} 1 & 2 & -1 \\ 4 & -3 & 4 \\ 2 & -1 & 1 \end{bmatrix} \mathbf{X} = \begin{Bmatrix} -3 \\ 1 \\ -2 \end{Bmatrix}$$

by using I3B3 to compute $\mathbf{A}^{-1}$ and then form the product $\mathbf{A}^{-1}\mathbf{B}$ to verify that $\mathbf{X}^T = [-2 \quad 1 \quad 3]$.
14. Write a program I2BY2, to invert a 2 by 2 matrix and determine its determinant.
15 Write a program, GETMAT, to extract **B** of order r x c from **A**, of order n x n. Let **B** start at A_{ij} and do not change **A**.

3. THE DIRECT APPROACH IN ONE-DIMENSION

3.1 Introduction

Clearly the simplest way to introduce finite element concepts is to begin in one dimension. Utilizing common physical concepts will allow us to postpone introducing various new mathematical concepts. Finally, the simplicity of one-dimensional elements lets the concepts of direct assembly of elements be illustrated in a heuristic manner. Selection of the common linear spring provides a good starting point.

3.2 The Linear Spring

The classical **linear spring** is well known to even freshman engineering students. Figure 3.1 illustrates a typical view of a spring. In that figure k denotes the stiffness, or modulus, of the spring, L is the unloaded length, F is the applied force, and d is the deformation of the spring. Recall that the spring was defined as linear if $F = kd$. There are two other related quantities that will be of importance to us. They are the **elastic potential energy**

$$U = \tfrac{1}{2} kd^2 \tag{3.1}$$

and the **work**, W, done by the constant applied force as it moves to the displaced point:

$$W = Fd. \tag{3.2}$$

Rather than consider a spring that always has one end fixed consider a generalized version as shown in Fig. 3.2. There x denotes a positive coordinate direction, and u_1 and u_2 are the displacements of the spring at ends 1 and 2, respectively. There is a corresponding set of nodal forces, F_1 and F_2. Noting that $d = (u_2 - u_1)$ the previous force-deformation law can be used to define the force at node 2:

$$F_2 = F = kd = k(u_2 - u_1) \tag{3.3}$$

and equilibrium requires that

$$F_1 = -F = -kd = -k(u_2 - u_1).$$

Reordering these equations gives

$$ku_1 - ku_2 = F_1$$

$$-ku_1 + ku_2 = F_2$$

which can be expressed in matrix form as

$$\begin{bmatrix} k & -k \\ -k & k \end{bmatrix} \begin{Bmatrix} u_1 \\ u_2 \end{Bmatrix} = \begin{Bmatrix} F_1 \\ F_2 \end{Bmatrix} \tag{3.4}$$

or symbolically, the **equilibrium equation** for the linear spring is

$$\mathbf{k}^e \mathbf{u}^e = \mathbf{f}^e.$$

The terminology that is used for these "element matrices" is that $\mathbf{k}^e$ is the element **stiffness matrix**, $\mathbf{f}^e$ is the element **force vector** and $\mathbf{u}^e$ is the vector of element **nodal displacements.** Usually the system of interest will consist of more than one spring element. In that case there would be an analogous, but different, set of "system matrices". Note that $\mathbf{k}^e$ is symmetric. It also has a zero determinant. Therefore, its inverse does not exist. That is because the equilibrium equation, Eq. (3.4), does not yet include a necessary geometric boundary condition.

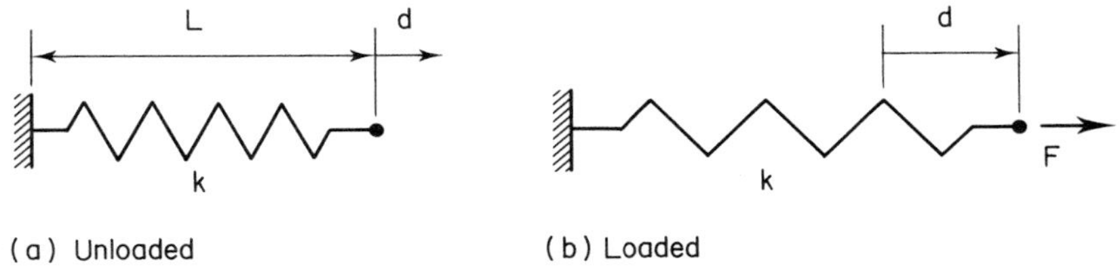

Fig. 3.1 A Linear Spring

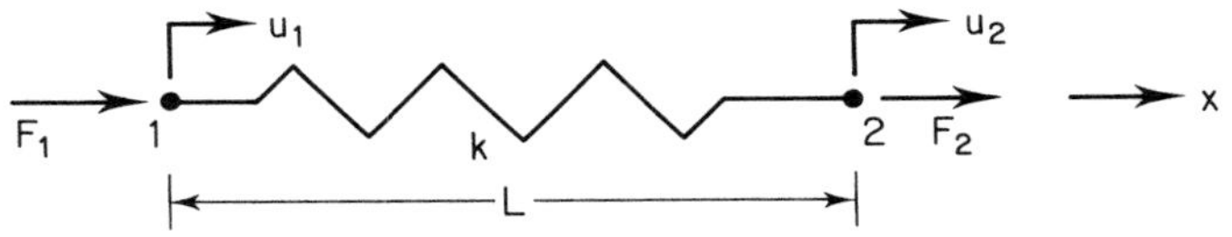

Fig. 3.2 A Generalized Spring Element

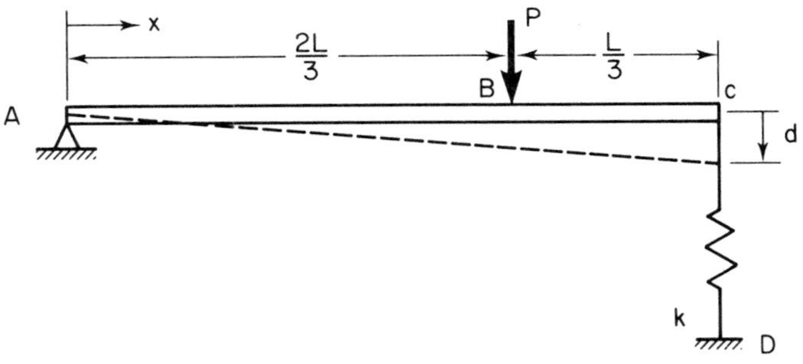

Fig. 3.3 A Spring-Linkage System

3.3 Energy Considerations

Before considering how to combine, or assemble, more than one element to form a system we will review some alternate views of equilibrium. Recall that most introductory texts on rigid body equilibrium (statics) usually employ the vector equations: $\Sigma \mathbf{F} = \mathbf{0}$, and $\Sigma \mathbf{M} = \mathbf{0}$. Such texts often discuss alternate formulations of equilibrium based on energy methods or virtual work. These alternate forms are easily extended to include elastic bodies, such as a spring.

One common approach considers a quantity called the **Total Potential Energy,** Π. This term is defined as the internal potential (elastic) energy, U, of the system minus the work of the externally applied forces, W. That is,

$$\Pi = U - W. \tag{3.5}$$

The **Theorem of Minimum Total Potential Energy** states: "The displacement field that satisfies the geometric boundary conditions and corresponds to the state of equilibrium is the one that minimizes the Total Potential Energy".

As an example of this theorem, return to the linear spring shown in Fig. 3.1. The geometric boundary condition, that the left end is fixed, is already satisfied. Thus the remaining question is what is the value of the displacement, d, where the spring reaches equilibrium with the applied constant force, F? Recalling our previous definitions of the strain energy and work of external forces gives

$$\Pi = {}^1\!/_2\, kd^2 - Fd. \tag{3.6}$$

This quantity depends only on the unknown displacement, d. To minimize Π we set its variation to zero. Thus

$$\frac{\partial \Pi}{\partial d} = 0 = \frac{2}{2}\, kd - F.$$

The resulting equilibrium equation is

$$kd = F \tag{3.7}$$

or d = F/k as expected. Next consider a similar problem shown in Fig. 3.3. Here the spring is again fixed at one end, D, and attached at C to a rigid link which is pinned at A. The link is loaded at an interior point, B, as shown in the figure. Again d is the unknown displacement. The potential energy of the spring is still $U = {}^1\!/_2\, kd^2$. To determine the work of the external force P it is necessary to know the displacement, d, at point C and be consistent with the geometric boundary conditions at A.

Since the linkage is rigid the displacement of any point on it can be obtained by linear interpolation. That is, $d(x) = (x/L)d_C$. Thus at point B, $x = 2L/3$, the displacement is $d_B = 2d_C/3$ and the work done by the force is

$$W = +2Pd_C/3$$

and the Total Potential Energy is

$$\Pi = kd_C^2/2 - 2Pd_C/3.$$

Minimizing Π with respect to d_C gives

$$0 = kd_C - 2P/3$$

as the equilibrium equation. Thus the spring deflection is $d_C = 2P/3k$ and the corresponding force in the spring is $F = kd_C = 2P/3$, as one would expect. The main point of this example is that one must often **interpolate** for displacements in order to be consistent with related calculations. For future reference note that the interpolation function could be generalized to

$$\begin{aligned} d(x) &= H_1(x)d_A + H_2(x)d_C \\ &= (1 - x/L)d_A + (x/L)d_C \end{aligned} \qquad (3.8)$$

to include the case where d_A is not zero.

One of our goals is to eventually automate these and related procedures. That is, the equilibrium equations can be generated, assembled, geometric boundary conditions introduced, displacements computed, and internal forces recovered by fairly simple computer algorithms. An important observation is that most of these procedures can be applied to other areas of analysis such as heat transfer, fluid flow, etc.

3.4 A Two Spring Assembly

Let us temporarily leave the concepts of energy methods and return to the direct assembly procedure for springs. Recall the matrix equilibrium equations, Eq. (3.4), for the single spring shown in Fig. 3.2. As we have seen, a typical spring has two nodes. When the system of interest consists of only one spring then the system has only two nodes. But in general the system will consist of several springs and several nodes. Thus it will become necessary to keep track of which two nodes in the system are attached to a particular element. This information is called the element connectivity or **topology.** Before establishing a precise algorithm for utilizing the topology we will introduce these concepts through heuristic arguments.

Consider a structural system which consists of two linear springs as shown in Fig. 3.4. We will refer to the system nodal displacements u_1, u_2, and u_3, as **u**. Two of these are associated with each element. Referring to Fig. 3.2 and Eq. (3.4) we can write the corresponding equilibrium equation for each element by inspection. That is,

$$\begin{bmatrix} k_a & -k_a \\ -k_a & k_a \end{bmatrix} \begin{Bmatrix} u_1 \\ u_2 \end{Bmatrix} = \begin{Bmatrix} F^a{}_1 \\ F^a{}_2 \end{Bmatrix} \qquad (3.9)$$

and

$$\begin{bmatrix} k_b & -k_b \\ -k_b & k_b \end{bmatrix} \begin{Bmatrix} u_2 \\ u_3 \end{Bmatrix} = \begin{Bmatrix} F^b{}_2 \\ F^b{}_3 \end{Bmatrix} . \qquad (3.10)$$

However, we are interested in the equilibrium of the system. Thus these element equations will be rewritten in terms of the system displacements $\mathbf{u}^T = [u_1 \;\; u_2 \;\; u_3]$. That can be done by adding a row and column of zeros to each of the above equation sets. Then

$$\begin{bmatrix} k_a & -k_a & 0 \\ -k_a & k_a & 0 \\ 0 & 0 & 0 \end{bmatrix} \begin{Bmatrix} u_1 \\ u_2 \\ u_3 \end{Bmatrix} = \begin{Bmatrix} F^1{}_a \\ F^2{}_a \\ 0 \end{Bmatrix}$$

and

$$\begin{bmatrix} 0 & 0 & 0 \\ 0 & k_b & -k_b \\ 0 & -k_b & k_b \end{bmatrix} \begin{Bmatrix} u_1 \\ u_2 \\ u_3 \end{Bmatrix} = \begin{Bmatrix} 0 \\ F^2{}_b \\ F^3{}_b \end{Bmatrix} .$$

These additional zeros are logical additions since they imply that element 'a' is unaffected by a displacement of node 3. Likewise, node 1 has no effect on element 'b'. Now the

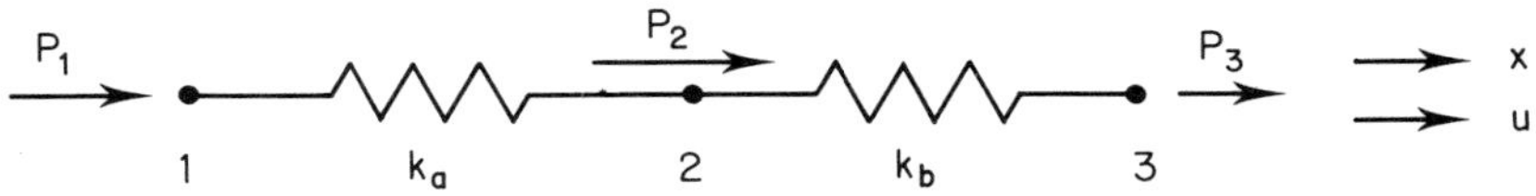

(a) Element assemblage

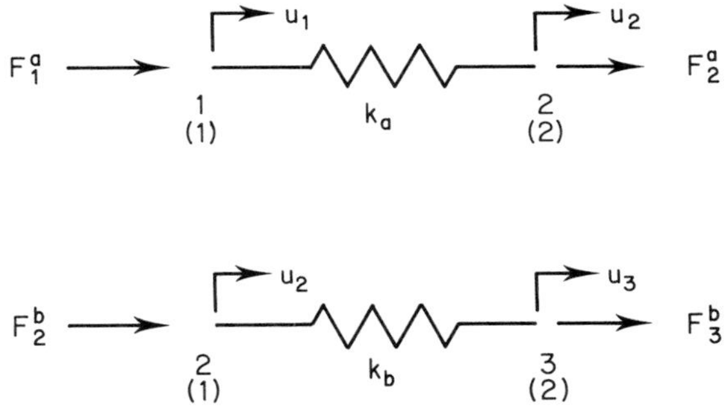

(b) Exploded view of elements

Element number	Topology 1	Topology 2
a	1	2
b	2	3

(c) Element topology lists

Fig. 3.4 A Two Element Spring Assembly

equations refer to the same variables and the matrices are of the same order. Thus we will superimpose the two equilibrium relations to obtain the system equilibrium equations:

$$\begin{bmatrix} k_a & -k_a & 0 \\ -k_a & (k_a+k_b) & -k_b \\ 0 & -k_b & k_b \end{bmatrix} \begin{Bmatrix} u_1 \\ u_2 \\ u_3 \end{Bmatrix} = \begin{Bmatrix} F^1{}_a \\ F^2{}_a + F^2{}_b \\ F^3{}_b \end{Bmatrix} = \begin{Bmatrix} P_1 \\ P_2 \\ P_3 \end{Bmatrix} \tag{3.11}$$

or symbolically the **system equilibrium** equations are

$$\mathbf{Ku} = \mathbf{P}. \tag{3.12}$$

This procedure is referred to as the **direct stiffness method** or direct analysis. Note that the right hand side of Eq. (3.11) represents the fact that at a typical system node, i, the externally applied force, P_i, equals the sum of the forces applied at the point by each of the elements connected to the point.

In Eq. (3.12) **K** and **P** are called the **system stiffness** and **system force vector**, respectively. Here **K** is symmetric and is currently singular since its determinant is zero. That is, $\mathbf{K}^{-1}$ does not exist since the system does not yet reflect any support specifications. Physically this means that the system could move (translate) as a rigid body. Therefore, to get a practical equilibrium state the geometric boundary conditions must be imposed on the problem.

3.5 Applying Essential Boundary Conditions

There are a number of procedures that can be utilized to impose essential boundary conditions. Matrix partitioning yields one popular algorithm. As a specific illustration of these concepts assume that u_3 in the previous example is specified to be zero. Note that the equilibrium equation could be partitioned as

$$\left[\begin{array}{cc:c} k_a & -k_a & 0 \\ -k_a & (k_a+k_b) & -k_b \\ \hdashline 0 & -k_b & k_b \end{array}\right] \left\{\begin{array}{c} u_1 \\ u_2 \\ \hdashline u_3 \end{array}\right\} = \left\{\begin{array}{c} P_1 \\ P_2 \\ \hdashline P_3 \end{array}\right\} . \tag{3.13}$$

Since u_3 is given (zero) this means that P_3 should be interpreted as an unknown **reaction.** Symbolically this partition can be written as

$$\begin{bmatrix} \mathbf{K}_{rr} & \mathbf{K}_{rg} \\ \mathbf{K}_{gr} & \mathbf{K}_{gg} \end{bmatrix} \begin{Bmatrix} \mathbf{u}_r \\ \mathbf{u}_g \end{Bmatrix} = \begin{Bmatrix} \mathbf{P}_r \\ \mathbf{P}_g \end{Bmatrix} \tag{3.14}$$

where the subscripts g and r are used to denote quantities related to given and required displacements, respectively. Notice that both $\mathbf{K}_{rr}$ and $\mathbf{K}_{gg}$ are square and symmetric and the rectangular partitions are related by $\mathbf{K}^T_{gr} = \mathbf{K}_{rg}$. It was mentioned earlier that the submatrices in the partitions can be treated algebraically like scalar matrix elements. Therefore, expanding the products in the first row from Eq. (3.14) gives

$$\mathbf{K}_{rr}\mathbf{u}_r + \mathbf{K}_{rg}\mathbf{u}_g = \mathbf{P}_r$$

or

$$\mathbf{K}_{rr}\mathbf{u}_r = \mathbf{P}_r - \mathbf{K}_{rg}\mathbf{u}_g \tag{3.15}$$

so that the required displacements are found from

$$\mathbf{u}_r = \mathbf{K}^{-1}{}_{rr}(\mathbf{P}_r - \mathbf{K}_{rg}\mathbf{u}_g) \tag{3.16}$$

since the partition $\mathbf{K}_{rr}$ is not singular even though $\mathbf{K}$ was singular. Often, as in our example, the given displacements are zero, i.e. $\mathbf{u}_g = \mathbf{0}$. This makes the results even simpler:

$$\mathbf{u}_r = \mathbf{K}^{-1}{}_{rr}\mathbf{P}_r \quad \text{for } \mathbf{u}_g = \mathbf{0}.$$

Once these required displacements are known the corresponding reactions, $\mathbf{P}_g$, can be recovered. They are obtained from the second row of Eq. (3.14)

$$\mathbf{P}_g = \mathbf{K}_{gr}\mathbf{u}_r + \mathbf{K}_{gg}\mathbf{u}_g. \tag{3.17}$$

To illustrate this conclusion consider a numerical example where the stiffnesses are k_a = 200N/m, k_b = 400 N/m, the externally applied forces are $\mathbf{P}_1$ = -30N, $\mathbf{P}_2$ = 20N, and the support condition is u_3 = 0. The system is shown in Fig. 3.5. Substituting into Eq. (3.11) gives

$$\begin{bmatrix} 200 & -200 & 0 \\ -200 & 600 & -400 \\ 0 & -400 & 400 \end{bmatrix} \begin{Bmatrix} u_1 \\ u_2 \\ 0 \end{Bmatrix} = \begin{Bmatrix} -30 \\ 20 \\ P_3 \end{Bmatrix}. \tag{3.18}$$

Noting that K_{gg} = [400], $K_{gr} = \mathbf{K}^T_{rg}$ = [0 -400], and $\mathbf{u}_g = \mathbf{0}$ Eq. (3.16) yields

$$\begin{Bmatrix} u_1 \\ u_2 \end{Bmatrix} = \begin{bmatrix} 200 & -200 \\ -200 & 600 \end{bmatrix}^{-1} \begin{Bmatrix} -30 \\ 20 \end{Bmatrix} + \begin{Bmatrix} 0 \\ 0 \end{Bmatrix}$$

$$\begin{Bmatrix} u_1 \\ u_2 \end{Bmatrix} = \frac{1}{80{,}000} \begin{bmatrix} 600 & 200 \\ 200 & 200 \end{bmatrix} \begin{Bmatrix} -30 \\ 20 \end{Bmatrix} = \begin{Bmatrix} -0.175 \\ -0.025 \end{Bmatrix}.$$

The deformation of the first spring is $d_a = (u_2 - u_1) = -0.025 - (-0.175) = +0.15$m and the force in it is $F_a = k_a d_a$ = (200 N/m)(0.15m) = +30N, tension. Similarly, d_b = +0.025m and F_b = +10N, tension. We could get the reaction at node 3 by inspection of the applied forces. However, other problems cannot be solved by inspection. Thus we compute the reaction from Eq. (3.17):

$$P_3 = P_g = [0 \quad -400] \begin{Bmatrix} -0.175 \\ -0.025 \end{Bmatrix} + [400]\ \{0\}$$

$$P_3 = +10\text{N}.$$

The positive sign of the answer shows that it is to the right as expected.

A minor inconvenience of the partition method is that it appears to require that all of the given system displacements need to be numbered so that they occur last in the list of system displacements. That is not actually required. The common alternative procedures can be suggested by considering the case where $\mathbf{u}_g$ is a single term. In that case Eq. (3.15) suggests that the reduced equations can be obtained by multiplying column g of **K** (i.e. partition $\mathbf{K}_{rg}$) by $\mathbf{u}_g$ and subtracting it from the external forces on the right. Then striking out the g-th column in **K** and the corresponding g-th row of **K** would give the reduced set. This procedure is popular for relatively small problems that are being solved by hand.

If the number of given terms, $\mathbf{u}_g$ is relatively small and the number of required terms, u_r, is relatively large it may not be convenient to reorder (renumber) the remaining equations after the rows and columns have been deleted. To keep the number of terms unchanged the deleted row could be replaced with the identity u_g = V where V is the assigned value of u_g. That is, a unity coefficient is placed on the diagonal of the g-th row of **K** and the g-th row of **P** is set to V.

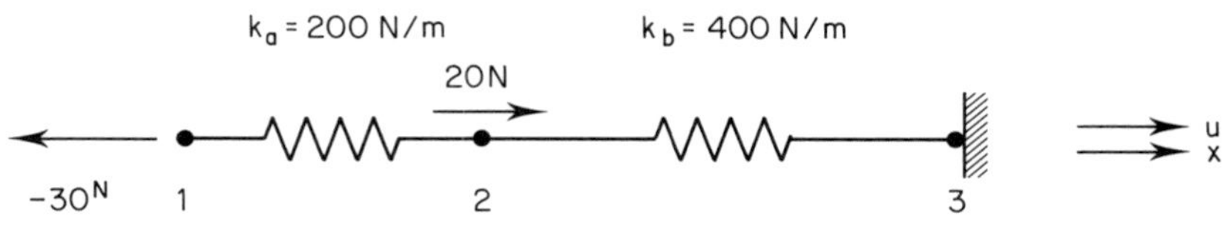

(a) Given conditions

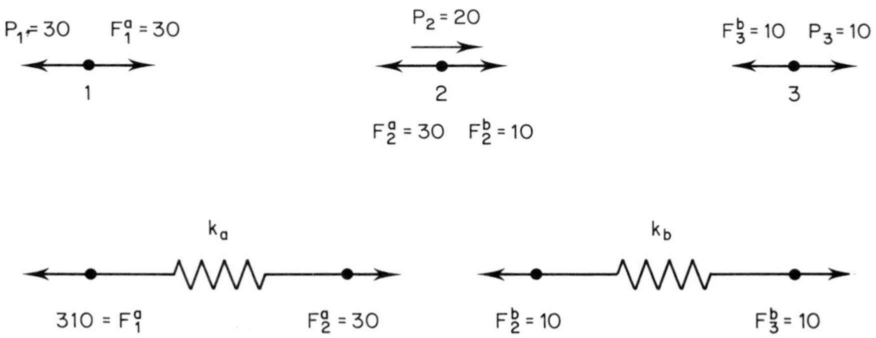

(b) Computed equilibrium

Fig. 3.5 Example Structural System

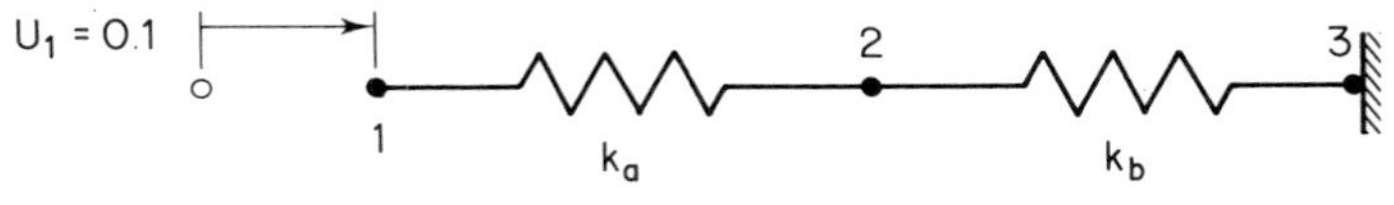

Fig. 3.6 An Indeterminate System

If the first alternative were used to include the boundary condition via a hand solution then the modified Eq. (3.18) would look like

$$\begin{bmatrix} 200 & -200 & 0 \\ -200 & 600 & -400 \\ 0 & -400 & 400 \end{bmatrix} \begin{Bmatrix} u_1 \\ u_2 \\ u_3 \end{Bmatrix} = \begin{Bmatrix} -30 \\ 20 \\ P_3 \end{Bmatrix} - 0 \begin{Bmatrix} 0 \\ -400 \\ 400 \end{Bmatrix} \qquad (3.19)$$

and the remaining two equations can be solved for u_1 and u_2. By way of comparison, if the second alternative modification was used the above equation would become

$$\begin{bmatrix} 200 & -200 & 0 \\ -200 & 600 & 0 \\ 0 & 0 & 1 \end{bmatrix} \begin{Bmatrix} u_1 \\ u_2 \\ u_3 \end{Bmatrix} = \begin{Bmatrix} -30 \\ 20 \\ 0 \end{Bmatrix}$$

and all three equations are solved. The solution would yield the identity that $u_3 = 0$. Clearly we solve more equations and lose reaction information, but we avoid reordering the matrix.

For very large problems applying the boundary conditions and solving the equations may represent a large percentage of the analysis cost. Thus these algorithms tend to be written very carefully and finely tuned for maximum efficiency.

However, the beginner need not be concerned with such refinements. For example, the operations in Eq. (3.16) could be carried out using subroutines like those given in the previous chapter, eg. MATMLT, MATSUB, and SYMINV. If one chooses this simplified approach, care must be used when extracting the partitions from **K**.

To illustrate a case where the reactions cannot be found by inspection consider a statically indeterminate system. Return to the system in Fig. 3.6. Instead of applying an external load at node 1 specify its displacement to be u_1 = +0.1m. The assembled equations are

$$\begin{bmatrix} 200 & -200 & 0 \\ -200 & 600 & -400 \\ 0 & -400 & 400 \end{bmatrix} \begin{Bmatrix} u_1 \\ u_2 \\ u_3 \end{Bmatrix} = \begin{Bmatrix} P_1 \\ 20 \\ P_3 \end{Bmatrix}. \qquad (3.20)$$

Modifying to include the two essential conditions gives

$$\begin{bmatrix} 200 & -200 & 0 \\ -200 & 600 & -400 \\ 0 & 400 & 400 \end{bmatrix} \begin{Bmatrix} u_1=0.1 \\ u_2 \\ u_3=0 \end{Bmatrix}$$

$$= \begin{Bmatrix} P_1 \\ 20 \\ P_3 \end{Bmatrix} - (+0.1) \begin{Bmatrix} 200 \\ -200 \\ 0 \end{Bmatrix} - (0) \begin{Bmatrix} 0 \\ -400 \\ 400 \end{Bmatrix}$$

or simply

$$600\ u_2 = 20 + 20 + 0 = 40$$

$$u_2 = 0.0667 \text{ m.}$$

Now backsubstituting the known **u** vector into Eq. (3.17) gives the reactions $P_1 = +6.667$ and $P_3 = -26.667$. Of course, $P_1 + P_2 + P_3 = 0$ as expected.

3.6 Analogous Systems

There are several engineering analysis problems that can be formulated in a manner that makes them analogous to the linear spring. A few of these systems will be rewritten in terms of our current notation. Some will be illustrated via numerical examples.

3.6.1 Heat Conduction

Consider the problem of one-dimensional steady state heat conduction. Recall **Fourier's law** that the heat flow rate, q, equals the product of the materials' thermal conductivity, K, the conducting cross-sectional area, A, and the temperature gradient, dT/dx, in the direction of conduction. That is,

$$q = -KA\frac{dT}{dx} \quad . \tag{3.21}$$

If we consider a homogeneous element of length $L = (x_2 - x_1)$ with a constant temperature gradient between its end temperatures then

$$\frac{dT}{dx} = \frac{T_2 - T_1}{x_2 - x_1} \quad .$$

Therefore in such an element

$$q^e = \frac{-K^e A^e}{L^e} (T_2^e - T_1^e) \quad . \tag{3.22}$$

If this is compared with Eq. (3.3) one notes the analogy: $F \sim q$, $k \sim KA/L$, and $u \sim T$. Using the analogy one can write the thermal equilibrium of an element, from Eq. (3.4), as

$$\frac{K^e A^e}{L^e} \begin{bmatrix} 1 & -1 \\ -1 & 1 \end{bmatrix} \begin{Bmatrix} T_1^e \\ T_2^e \end{Bmatrix} = \begin{Bmatrix} q_1^e \\ q_2^e \end{Bmatrix}$$

or symbolically

$$\mathbf{K}^e \mathbf{T}^e = \mathbf{q}^e \ .$$

If these element relations are assembled to describe a thermal system the result is

$$\mathbf{KT} = \mathbf{Q} \tag{3.23}$$

where the right hand side is the externally applied nodal heat flux. It reflects the continuity or conservation of the nodal heat flow so at a typical node, say i,

$$Q_i = \sum_e q_i^e \ . \tag{3.24}$$

In other words, the sum of the heat flows in all elements connected to node i equals the external heat flux at that node.

Note that by assemblying piecewise homogeneous elements that have different conductivities it is possible to treat a non-homogeneous problem. This is a standard feature, and major advantage, of the finite element method. As an example, consider a furnace wall as shown in Fig. 3.7. It consists of 9 inches of firebrick (K_1 = 0.72 BTU/(hr ft F), 5 inches of insulation brick (K_2 = 0.08), and 7.5 in. of red brick (K_3 = 0.5). The specified inner and outer temperatures are 1500 F and 150 F, respectively. We wish to determine the internal temperature distribution.

This problem can be treated in a one-dimensional model. A constant arbitrary heat conduction area of A = 1 sq. ft. is assumed for all elements. A system with four nodes and three elements is selected. A summary of the properties is

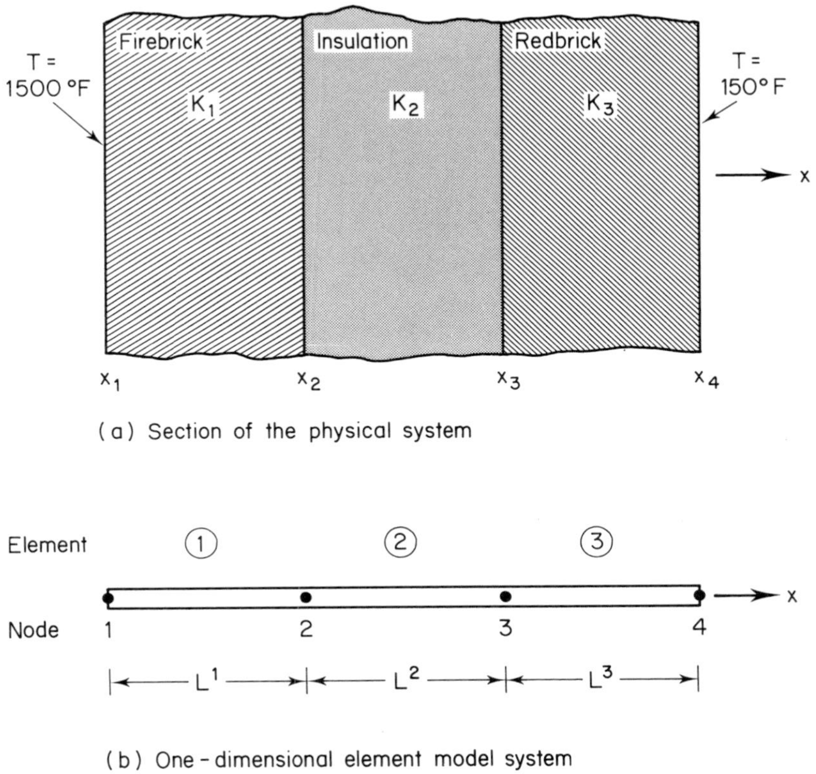

Fig. 3.7 Steady State Heat Conduction in a Wall

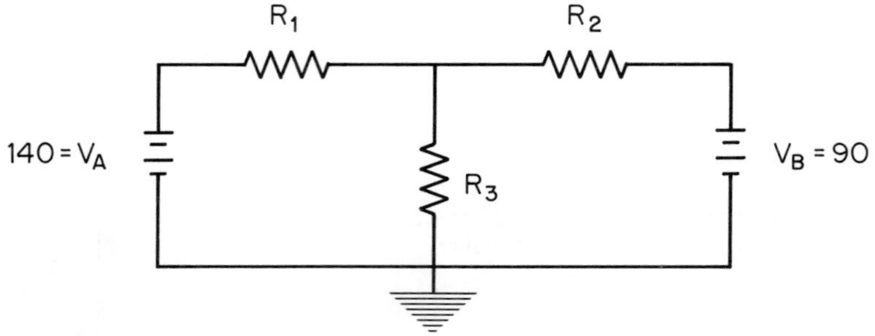

(a) A typical DC circuit

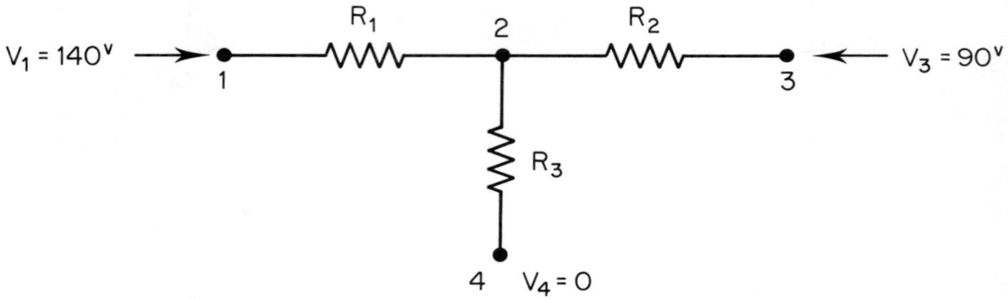

(b) A finite element mesh

Element, e	R^e	Topology
1	20 ohms	1, 2
2	5 ohms	2, 3
3	6 ohms	2, 4

Essential conditions : $V_1 = 140$, $V_3 = 90$, $V_4 = 0$ volts

(c) System properties

Fig. 3.8 A DC Circuit Model

Element, e	A^e	L^e	K^e	Topology
1	1	9/12	0.72	1,2
2	1	5/12	0.08	2,3
3	1	7.5/12	0.50	3,4

and the essential boundary conditions are $T_1 = 1500$ and $T_4 = 150$. The assembled system is

$$\frac{1(12)}{100}\begin{bmatrix} 72/9 & -72/9 & 0 & 0 \\ -72/9 & (72/9+8/5) & -8/5 & 0 \\ 0 & -8/5 & (8/5+50/7.5) & -50/7.5 \\ 0 & 0 & -50/7.5 & -50/7.5 \end{bmatrix}\begin{Bmatrix} T_1 \\ T_2 \\ T_3 \\ T_4 \end{Bmatrix} = \begin{Bmatrix} Q_1 \\ 0 \\ 0 \\ Q_4 \end{Bmatrix}. \qquad (3.25)$$

Solving gives $T_2 = 1312.50$ F and $T_3 = 375.00$ F. Backsubstitutions of the temperatures yields the thermal "reaction": $Q_1 = 180.00$ BTU/hr $= Q_4$. Thus the heat flow is in the positive x direction.

3.6.2 Electrical Resistance Networks

Consider an electric resistance element connecting two nodes in a DC circuit. **Ohm's law** gives the relation between the direction current, j, entering the element, the voltage drop $(E_2 - E_1)$, and the resistance, R, of the material. Specifically

$$j = \frac{1}{R}(E_2 - E_1)$$

so by analogy: $F \sim j$, $k \sim 1/R$, $u \sim E$, and

$$\frac{1}{R^e}\begin{bmatrix} 1 & -1 \\ -1 & 1 \end{bmatrix}\begin{Bmatrix} E_1^e \\ E_2^e \end{Bmatrix} = \begin{Bmatrix} j_1^e \\ j_2^e \end{Bmatrix}. \qquad (3.26)$$

Symbolically this element relation is $\mathbf{K}^e\mathbf{E}^e = \mathbf{j}^e$ and the corresponding system relationship is $\mathbf{KE} = \mathbf{J}$. Here again $\mathbf{J}$ is the vector of external nodal currents. That is, at each node J equals the sum of the currents, j, from the connecting elements.

As an example, consider the DC circuit illustrated in Fig. 3.8. Assembling the system shown in Fig. 3.6 (b) gives

$$\begin{bmatrix} 1/20 & -1/20 & 0 & 0 \\ -1/20 & (1/20+1/5+1/6) & -1/5 & -1/6 \\ 0 & -1/5 & 1/5 & 0 \\ 0 & -1/6 & 0 & 1/6 \end{bmatrix} \begin{Bmatrix} E_1=140 \\ E_2 \\ E_3=90 \\ E_4=0 \end{Bmatrix} = \begin{Bmatrix} J_1 \\ 0 \\ J_3 \\ J_4 \end{Bmatrix}.$$

Solving yields $(1/20 + 1/5 + 1/6)\ E_2 = 0 + 140/20 + 90/5$ or simply $E_2 = 60$ volts. Substituting the voltages to determine the 'reaction' currents gives $J_1 = 4$ amps, $J_3 = 6$, and $J_4 = -10$. That is, external current entered the system at nodes 1 and 3 and was removed at node 4. Post-processing the results gives the current in each element. The results, for the first node in the topology, are

$$j_1 = (E_2 - E_1)/R_1 = (60 - 140)/20 = -4, \text{ entering}$$

$$j_2 = (E_3 - E_2)/R_2 = (90 - 60)/5 = +6, \text{ exiting}$$

$$j_3 = (E_4 - E_2)/R_3 = (0 - 60)/6 = -10, \text{ entering} \quad .$$

In a similar manner the power $P = EI = I^2R$ can be computed. The external power is $P = \mathbf{E}^T\mathbf{J} = (140)4 + 0 + (90)6 + 0 = 1100$ watts and the internal power is the sum of the power in the elements. That is,

$$P = \sum_e E^{eT}\ \mathbf{K}^e E^e$$

$$P_1 = \begin{matrix} 140 & 60 \end{matrix} \begin{bmatrix} 1/20 & -1/20 \\ -1/20 & 1/20 \end{bmatrix} \begin{Bmatrix} 140 \\ 60 \end{Bmatrix} = 320$$

$$P_2 = \begin{matrix} 60 & 90 \end{matrix} \begin{bmatrix} 1/5 & -1/5 \\ -1/5 & 1/5 \end{bmatrix} \begin{Bmatrix} 60 \\ 90 \end{Bmatrix} = 180$$

$$P_3 = \begin{matrix} 60 & 0 \end{matrix} \begin{bmatrix} 1/6 & -1/6 \\ -1/6 & 1/6 \end{bmatrix} \begin{Bmatrix} 60 \\ 0 \end{Bmatrix} = 600$$

and $P = 320 + 180 + 600 = 1100$ watts, as expected. These procedures can be used on DC systems in general, but it can become difficult to clarify the topology and the boundary conditions.

3.6.3 A Truss Element

A truss element transmits a force between its two end points. This is equivalent to the uniaxial bar considered in a course on the strength of materials. Recall that a bar of material modulus E, cross-sectional area, A, and length, L, will elongate when subjected to an axial end force F, and fixed at the other end. The deformation is given in such courses as:

$$U = \frac{FL}{AE} \quad \text{so } F = \left(\frac{AE}{L}\right) u \ . \tag{3.26}$$

This suggests the analogy that $F \sim F$, $k \sim (AE/L)$, and $u \sim u$. Thus

$$\frac{A^e E^e}{L^e} \begin{bmatrix} 1 & -1 \\ -1 & 1 \end{bmatrix} \begin{Bmatrix} u_1^e \\ u_2^e \end{Bmatrix} = \begin{Bmatrix} F_1^e \\ F_2^e \end{Bmatrix} . \tag{3.27}$$

As noted in Fig. (3.9), it is not uncommon for a bar element to have an internal load contributing to a set of statically equivalent nodal forces on the element. In this case the equivalent forces are simply half the weight lumped to each end of the element.

3.6.4 A Torsion Element

Another common problem in the study of strength of materials is the torsion of a circular rod. Consider a rod fixed at one end and subjected to an axial torque, T, at the other end. Let the length, shear modulus of the material, and the polar moment of inertia of the cross-section be denoted by L, G, and J, respectively. Then the resulting rotation, in radians, is illustrated in Fig. 3.10. Its value is

$$\Theta = \frac{TL}{JG} \quad \text{or } T = \left(\frac{JG}{L}\right)\Theta \ .$$

The torsional analogy is

$$\frac{J^e G^e}{L^e} \begin{bmatrix} 1 & -1 \\ -1 & 1 \end{bmatrix} \begin{Bmatrix} \theta_1^e \\ \theta_2^e \end{Bmatrix} = \begin{Bmatrix} T_1^e \\ T_2^e \end{Bmatrix} . \tag{3.28}$$

As an example of the use of the torsional element consider the straight homogeneous circular bar shown in Fig. 3.10. The statically indeterminate system is fixed at each end. It has torques of $-T_A$ and T_B applied at its second and third points. The nodes and elements are numbered consecutively from the left end. Since the element properties are all the same they are factored out of the assembled system equations:

$$\frac{3JG}{L}\begin{bmatrix} 1 & -1 & 0 & 0 \\ -1 & 2 & -1 & 0 \\ 0 & -1 & 2 & -1 \\ 0 & 0 & -1 & 1 \end{bmatrix}\begin{Bmatrix} \Theta_1 \\ \Theta_2 \\ \Theta_3 \\ \Theta_4 \end{Bmatrix} = \begin{Bmatrix} T_L \\ -T_A \\ T_B \\ T_R \end{Bmatrix} \qquad (3.29)$$

where $T_L = T_1$ and $T_R = T_4$ are the unknown reactions at the left and right ends, respectively. Applying the boundary conditions that $\Theta_1 = 0 = \Theta_4$ gives

$$\frac{3JG}{L}\begin{bmatrix} 2 & -1 \\ -1 & 2 \end{bmatrix}\begin{Bmatrix} \theta_2 \\ \theta_3 \end{Bmatrix} = \begin{Bmatrix} -T_A \\ T_B \end{Bmatrix} \qquad (3.30)$$

inverting the square matrix gives

$$\begin{Bmatrix} \theta_2 \\ \theta_3 \end{Bmatrix} = \frac{L}{9JG}\begin{bmatrix} 2 & 1 \\ 1 & 2 \end{bmatrix}\begin{Bmatrix} -T_A \\ T_B \end{Bmatrix} = \frac{L}{9JG}\begin{Bmatrix} T_B-2T_A \\ 2T_B-T_A \end{Bmatrix}. \qquad (3.31)$$

Substituting these rotations into the first equation gives the left reaction:

$$\frac{3JG}{L}\left(0 - \frac{L}{9JG}(T_B - 2T_A) + 0 + 0\right) = (2T_A - T_B)/3 = T_L$$

and the last equation yields T_R:

$$\frac{3JG}{L}\left(0 + 0 - \frac{L}{9JG}(2T_B - T_A) + 0\right) = (T_A - 2T_B)/3 = T_R$$

as expected from mechanics of materials.

3.6.5 Porous Media Flow

The flow of ground water is governed by a principal known as **Darcy's law.** It states that the flow rate, v, through the porous media is proportional to the head loss and inversely proportional to the length of the flow path. If x is the direction of flow and h is the hydraulic head then Darcy's law is

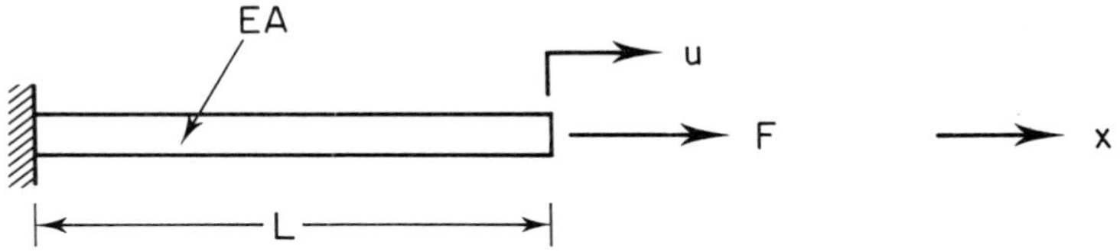

(a) A typical bar

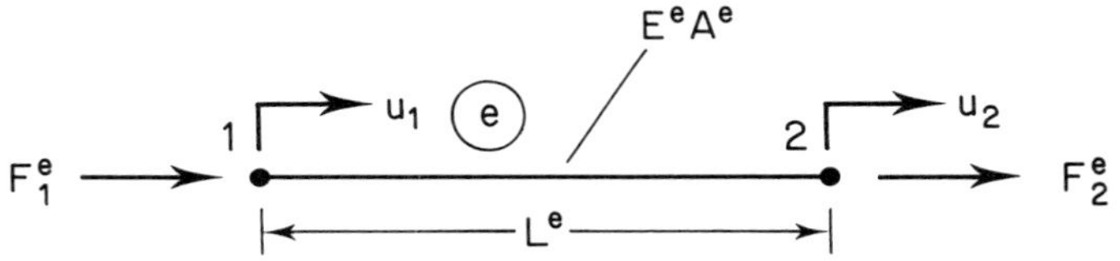

(b) A general bar element

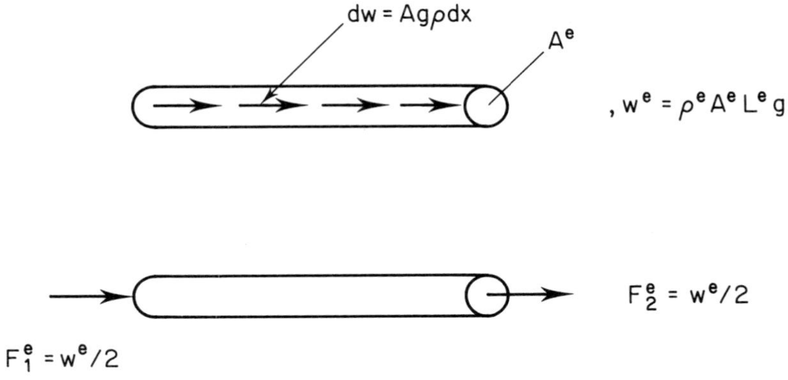

(c) Consistent nodal loads for axial gravity

Fig. 3.9 A Truss or Bar Element

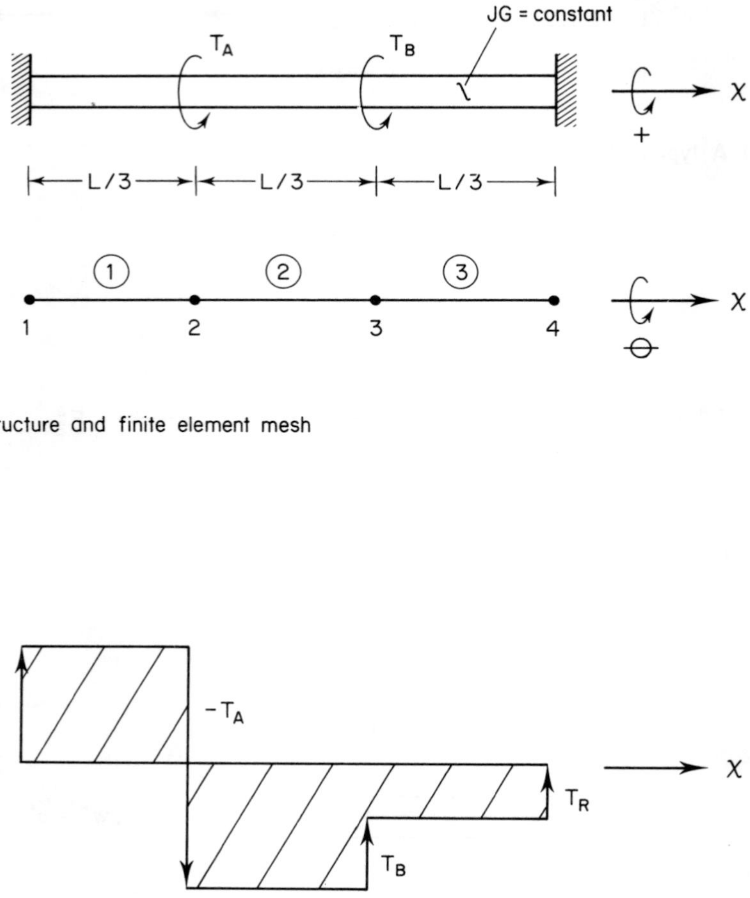

(a) Structure and finite element mesh

(b) Torque versus position

Fig. 3.10 An Elastic Torsion Element

$$v = -k\frac{dh}{dx} . \tag{3.32}$$

Here k is the coefficient of permeability. This can be compared directly with the heat conduction equation, Eq. (3.21). This would allow the finite element formulation of porous media to be written by inspection.

3.6.6 Laminar Pipe Flow

Consider the laminar flow in a circular pipe. The volumetric flow rate, q_1, is related to the pressure drop ($P_1 - P_2$) between points 1 and 2 that are a distance L apart by

$$q = R(P_1 - P_2) .$$

Here R denotes the flow resistance of the pipe and is given by $R = \pi D^4/(128\ L\mu)$ where D is the diameter and μ is the viscosity of the fluid. The fluid entering at node 2 is $q_2 = -q_1 = R(P_2 - P_1)$. If we express these identities in matrix form

$$\frac{\pi D^4}{128 L\mu}\begin{bmatrix} 1 & -1 \\ -1 & 1 \end{bmatrix}\begin{Bmatrix} P_1 \\ P_2 \end{Bmatrix} = \begin{Bmatrix} q_1 \\ q_2 \end{Bmatrix} \tag{3.33}$$

or

$$K^e P^e = q^e .$$

The corresponding system relations are **KP** = **Q** where **Q** is the vector of external input flows. Thus at any point **Q** equals the sum of the $\mathbf{q}^e$. Figure 3.11 shows the typical pipe flow element and the typical porous media element.

3.7 Automated Assembly and Essential Condition Modifications

Previously we have seen that the system matrices **K** and **P** were initially zero and then we added the terms from each element matrix, i.e. $\mathbf{K}^e$ and $\mathbf{P}^e$. The matrices **K** and **P** relate to the system displacements, or system **degrees of freedom, u.** By way of comparison $\mathbf{K}^e$ and $\mathbf{P}^e$ are related to the element displacements, or degrees of freedom, $\mathbf{u}^e$. The element quantities $\mathbf{u}^e$ are a partition or subset of **u.** It is the element topology, or connectivity list, that guides one in extracting the $\mathbf{u}^e$ from the **u.** It also allows one to relate the row and column subscripts of $\mathbf{K}^e$ or $\mathbf{P}^e$ to the corresponding terms in **K** or **P.**

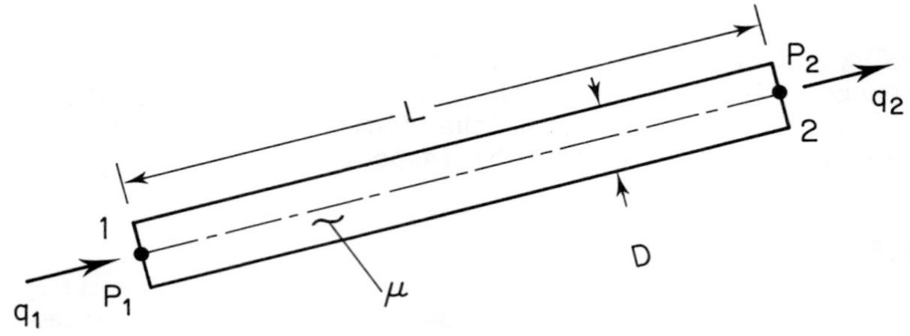

(a) A laminar flow pipe element

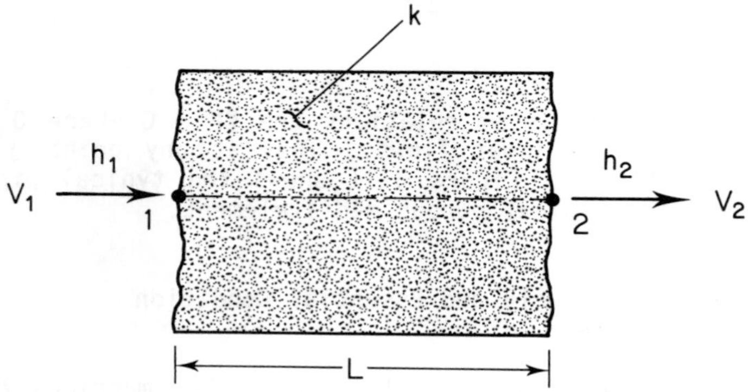

(b) A porous media flow element

Fig. 3.11 Typical one-dimensional flow elements

That is, for any typical element, e, each term in $\mathbf{K}^e$ or $\mathbf{P}^e$ is added into the system equations. However, we can only add matrices of the same order, and $\mathbf{K}$ and $\mathbf{K}^e$ are generally of different orders. In sec. 3.5 rows and columns of zero were added to $\mathbf{K}^e$ (or $\mathbf{P}^e$) until it was the same order as $\mathbf{K}$ (or $\mathbf{P}$). That operation in effect changed the row and column numbers of typical non-zero terms in $\mathbf{K}^e$ and $\mathbf{P}^e$ so that they could be directly added to $\mathbf{K}$ and $\mathbf{P}$.

We would like to automate the procedure of changing the element subscsripts and executing the addition. Thus for any element, e, we will form new coefficients in $\mathbf{K}$ or $\mathbf{P}$ from $\mathbf{K}^e$ or $\mathbf{P}^e$ by computing

$$P_I = P_I + p_i^e \qquad \forall\ i \tag{3.34}$$

$$K_{I,J} = K_{I,J} + k_{i,j}^e, \qquad \forall\ i \text{ and } j. \tag{3.35}$$

Here i and j are the local subscripts of a coefficient in an element matrix, and I, J are the corresponding subscripts of the system equation coefficient to which the element contribution is to be added. The direct subscript conversions for element e are defined as

$$\begin{aligned} I &= \text{INDEX } (i) \\ J &= \text{INDEX } (j) \end{aligned} \tag{3.36}$$

where the array INDEX is computed from the element topology list. If, as in our current examples, there is only one unknown per node then the array INDEX corresponds directly to the element topology list. Generally a slightly more complicated relationship exists. Since the array INDEX can be easily defined the assembly procedure can be automated. A subroutine, ADDCOL, for assemblying the force vector is shown in Fig. 3.12. A similar routine, ADDSQ, for assemblying the square matrices, is given in Fig. 3.13. To avoid application specific language these subroutines simply refer to the element square and column matrices, $\mathbf{S}^e$ and $\mathbf{C}^e$.

```
      SUBROUTINE ADDCOL (NSF,NLF,INDEX,C,CC)
C-->  ADD ELEMENT COLUMN MATRIX TO SYSTEM COLUMN
      DIMENSION C(NLF), CC(NSF), INDEX(NLF)
C     NSF = NO.OF SYS. DOF, NLF = NO. ELEM. DOF
      DO 10 I = 1,NLF
      II = INDEX(I)
10    CC(II) = CC(II) + C(I)
      RETURN
      END
```

Fig. 3.12 Addition of element column matrix to system column

```
      SUBROUTINE ADDSQ (NSF,NLF,INDEX,S,SS)
C-->  ADD ELEMENT SQ. MATRIX TO SYSTEM SQUARE
      DIMENSION S(NLF,NLF), SS(NSF,NSF), INDEX(NLF)
C     NSF = NO.OF SYS. DOF, NLF = NO. ELEM. DOF
      DO 20 J = 1,NLF
      JJ = INDEX(J)
      DO 10 I = 1,NLF
      II = INDEX(I)
10    SS(II,JJ) = SS(II,JJ) + S(K,J)
20    CONTINUE
      RETURN
      END
```

Fig. 3.13 Addition of element square matrix into system matrix

The concept of subscript conversion and the superposition of terms from various elements is graphically illustrated in Fig. 3.14 for a one dimensional problem. To illustrate the concept let us repeat the assembly for the problem shown in Fig. 3.5. First we zero **K** and **P**. Next P is set equal to the external loads. Then we tabulate INDEX for each element by using the element topology:

Element Number	Element Subscript, i	System Subscript or INDEX (i)
a	1	1
a	2	2
b	1	2
b	2	3

From symmetry we will only consider the upper triangle of $\mathbf{K}^e$. For each element we get:

Element (a) $\quad k_{11}^a: \; K_{11} = K_{11} + k_{11}^a = 0 + 200$

$k_{12}^a: \; K_{12} = K_{12} + k_{12}^a = 0 - 200$

$k_{22}^a: \; K_{22} = K_{22} + k_{22}^a = 0 + 200$

Element (b) $\quad k_{11}^b: \; K_{22} = K_{22} + k_{11}^b = 200 + 400$

$K_{12}^b: \; K_{23} = K_{23} + k_{12}^b = 0 - 400$

$K_{22}^b: \; K_{33} = K_{33} + k_{22}^b = 0 + 400$

so the assembled system **K** is:

$$\mathbf{K} = \begin{bmatrix} 200 & -200 & 0 \\ & 600 & -400 \\ \text{sym} & & 400 \end{bmatrix} .$$

Since all the $\mathbf{p}^e = \mathbf{0}$ there is no need to assemble additional contributions to **P**. So **P** remains at its initial value, $\mathbf{P}^T = [0 \quad 20 \quad 0]$.

The above assembly can be easily generalized to the case where there is more than one degree of freedom per node. Let there be NG generalized degrees of freedom at every node in the system. Then the degree of freedom number, say NDOF, of the j-th degree of freedom at node i is (by induction)

$$\text{NDOF} = \text{NG}(i - 1) + j \tag{3.37}$$

where $1 \leq j \leq \text{NG}$. Equation (3.37) can be used to compute the element or system equation numbers (subscripts). That is, Eq. (3.37) defines the terms in the array INDEX. Let the system have a total of M nodes and assume an element has N nodes. Note that $M \geq N$. To get the element dof numbers one would let i range over the local element nodal numbering, $1 \leq i \leq N$. Similarly to get all the system dof numbers one would use the range $1 \leq i \leq M$.

To carry out the actual direct assembly process it is necessary to find the system subscript that corresponds to an element subscript. These only require that we find the system node number that corresponds to the local element node number. That information is always given as input in the element topology or connectivity lists. The above bookkeeping procedure is illustrated for a series of one-dimensional elements in Fig. 3.15 where a system with four line elements, five nodes and three degrees of freedom (dof) per node is illustrated (i.e. M = 5, NE = 4, NG = 3, N = 2). There are a total of fifteen dof in the system. We wish to determine the dof number of the third parameter (J = 3) at system node number four (I = 4). Equation (3.37) shows that the required result is NDOF = 3(4 - 1) + 3 = 12 for the system dof number. For element three this corresponds to local dof number 6 while for element four it is local dof number 3. Therefore, we note that contributions to system equation number 12 comes from parts of two different element equation sets.

TABLE III.I Relating Element and System Equations

Local node i	Parameter number J	System node I=LNODE(i)	Equation Number Local: NG*(i-1)+J	Equation Number System NG*(I-1)+J
1	1	LNODE(1)	1	NG*(LNODE(1) - 1) + 1
1	2	LNODE(1)	2	
.	.	.		
.	.	.	.	.
.	.	.	.	
1	NG	LNODE(1)	.	.
2	1	LNODE(2)	.	.
.	.	.		
.	.	.		
.	.	.		
K	J	LNODE(K)	NG*(K - 1) + J	NG*(LNODE(K) - 1) + J
.	.	.		
.	.	.		
.	.	.		
N	1	LNODE(N)	.	.
.	.	.	.	.
.	.	.		
N	NG	LNODE(N)	N*NG	NG*(LNODE(N) - 1) + NG

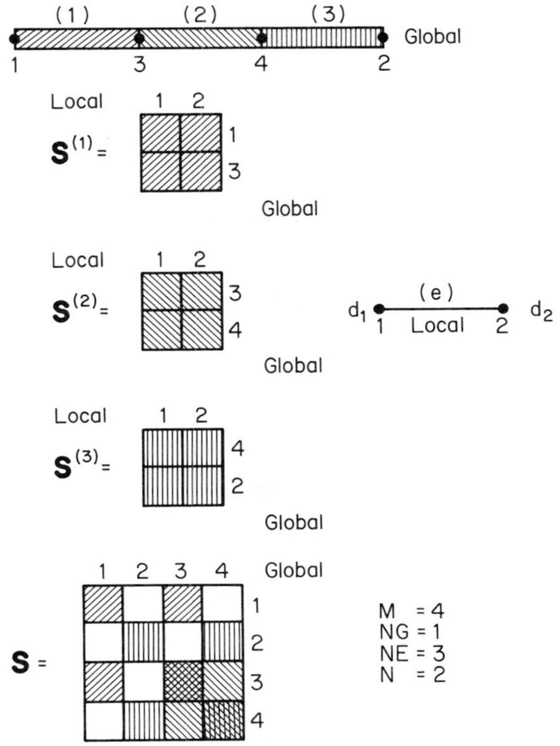

Fig. 3.14 Graphical Assembly For A Line Element With One Parameter Per Node

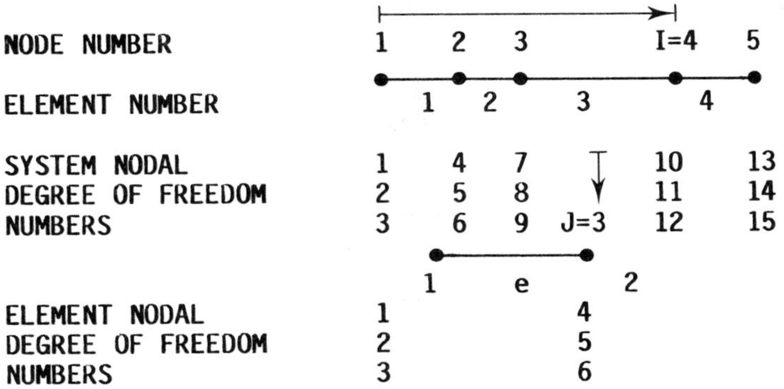

Fig. 3.15 Calculation of Degree of Freedom Numbers

By utilizing the list of N nodal points (element incidences) associated with a specific element, it is possible to again apply Eq. (3.37) and calculate the system degree of freedom numbers that correspond to the N*NG parameters assigned to the specific element. These calculations, illustrated in Table III.I are executed by subroutine INDXEL as shown in Fig. 3.16. The procedure for extracting the required element topology from the given topology data is illustrated in Fig. 3.17.

```
      SUBROUTINE INDXEL (N,NG,NLF,LNODE,INDEX)
C-->  FIND DOF NUMBERS FOR AN ELEMENT
      DIMENSION INDEX(NLF), LNODE(N)
C     N=NO. NODES PER ELEM, NG=NO. DOF PER NODE
C     NLF=N*NG=NO. ELEM DOF,
      DO 20  I = 1,N
      DO 10  J = 1,NG
      IJL = NG*(I - 1) + J
      IJS = NG*(LNODE(I) - 1) + J
10    INDEX(IJL) = IJS
20    CONTINUE
      RETURN
      END
```

Fig. 3.16 Computing Element Degree of Freedom Numbers

A numerical example of computing the dof numbers and of their use in assembly will be considered here. Assume that the typical element has three nodes (N = 3) and two dof per node (NG = 2). For example, this may be a beam with two end nodes and one midpoint node. The degrees of freedom could be the deflection (J = 1) at the node and the rotation (J = 2). Assume the topology data to be

System	261	270	310	(Array LNODE)
Local	1	2	3	.

Therefore, we find the degree of freedom numbers for the element as given in Table III.II

```
      SUBROUTINE LNODES (L,NE,N,NODES,LNODE)
C--   EXTRACT TOPOLOGY FOR ELEMENT L FROM NODES
      DIMENSION NODES(NE,N), LNODE(N)
C     NE=NO. ELEMS, N=NO. NODES PER ELEMENT
C     NODES=GIVEN TOPOLOGY DATA
      DO 10  J = 1,N
10    LNODE(J) = NODES(L,J)
      RETURN
      END
```

Fig. 3.17 Extracting Nodes on an Element

TABLE III.II Example Equation Number Conversion

Node local i	Number: system* I	Parameter number: J	Degree of freedom: system** NDOF	local NDOF
1	261	1	521	1
		2	522	2
2	270	1	539	3
		2	540	4
3	310	1	619	5
		2	620	6

* Input data array LNODE
** Returned conversion array INDEX

Recall that the element equations are expressed in terms of local degree of freedom numbers. In order to add the element coefficients into the system equations one must identify the relation between the local degree of freedom numbers and the corresponding system degree of freedom numbers. Array INDEX provides this information for a specific element.

In practice, the assembly procedure is as follows. First the system matrices are set equal to zero. Then a loop over all the elements is performed. For each element, the element matrices are generated in terms of the local degrees of freedom. The coefficients of the element matrices are added to the corresponding coefficients in the system matrices. Before the addition is carried out, the element array INDEX is used to convert the local subscripts of the coefficients to the system subscripts of the term in the system equations to which the coefficient is to be added. That is,

$$S^e_{i,j} \rightarrow S_{I,J}$$

$$C^e_i \rightarrow C_I$$

where I = INDEX(i) and J = INDEX(j) are the corresponding row and column numbers in the system equations; i,j are the subscripts of the coefficients in terms of the local degrees of freedom, and $\rightarrow$ reads "is added to." Considering the coefficients in the element matrices for the element in the current example, one finds

$$S^e_{1,1} \rightarrow S_{521,521} \quad C^e_1 \rightarrow C_{521}$$

$$S^e_{2,3} \rightarrow S_{522,539} \quad C^e_2 \rightarrow C_{522}$$

$$S^e_{3,4} \rightarrow S_{539,540} \quad C^e_3 \rightarrow C_{539}$$

$$S^e_{4,5} \rightarrow S_{540,619} \quad C^e_4 \rightarrow C_{540}$$

$$S^e_{5,6} \rightarrow S_{619,620} \quad C^e_5 \rightarrow C_{619}$$

$$S^e_{1,6} \rightarrow S_{521,620} \quad C^e_6 \rightarrow C_{620}, \text{ etc.}$$

The procedure for modifying the system matrices to include specified dof can also be automated. A popular method is to modify **P** using the columns of **K** and then to insert an identity on a row to avoid renumbering the equations. One such algorithm, MODIFY, is shown in Fig. 3.18 and the resulting changes are shown in Fig. 3.19. It is called once for each dof that has a given value.

Nodal parameter constraints can also be accomplished by the use of an artifice such as the following. Let 'i' denote the degree of freedom to be assigned a given value, say V. This trick consists of modifying two terms of the system matrix and column vector, i.e., $S_{i,i}$, and C_i. These terms are redefined to be

$$C_i = bVS_{i,i}$$

$$S_{i,i} = bS_{i,i}$$

where b is a very large number. This yields the i-th system equation

$$S_{i,1}D_1 + \dots + bS_{i,i}D_i + \dots S_{i,n}D_n = bVS_{i,i}$$

which is a good approximation of the boundary condition $D_i = V$, if b is sufficiently large. This artifice is the fastest method for introducing constraints but it can lead to numerical ill-conditioning. There is also a problem of how large a value should be assigned to b. Commonly used values range from ten to the 12-th power, to the 28-th power.

```
      SUBROUTINE MODIFY (NSF,J,VALUE,S,C)
C-->  APPLY B.C. TO A SET OF SYMMETRIC EQUATIONS
C     S*D=C, AND D(J)=VALUE
      DIMENSION  S(NSF,NSF), C(NSF)
C     NSF=NO. OF DOF IN EQUATIONS
      DO 10 I = 1,NSF
      C(I) = C(I) - VALUE*S(I,J)
      S(I,J) = 0.0
 10   S(J,I) = 0.0
      S(J,J) = 1.0
      C(J) = VALUE
      RETURN
      END
```

Fig. 3.18 A Row and Column Modification Algorithm

$$\begin{bmatrix} S_{11} & \cdots & 0 & \cdots & S_{1k} & \cdots & S_{1p} \\ \vdots & & \vdots & & \vdots & & \vdots \\ 0 & \cdots & 1 & \cdots & 0 & \cdots & 0 \\ \vdots & & \vdots & & \vdots & & \vdots \\ S_{k1} & \cdots & 0 & \cdots & S_{kk} & \cdots & S_{kp} \\ \vdots & & \vdots & & \vdots & & \vdots \\ S_{p1} & \cdots & 0 & \cdots & S_{pk} & \cdots & S_{pp} \end{bmatrix} \begin{Bmatrix} D_1 \\ \vdots \\ D_j \\ \vdots \\ D_k \\ \vdots \\ D_p \end{Bmatrix} = \begin{Bmatrix} C_1 - bS_{1j} \\ \vdots \\ b \\ \vdots \\ C_k - bS_{kj} \\ \vdots \\ C_p - bS_{pj} \end{Bmatrix}$$

Fig. 3.19 A Modification to Include Forced Boundary Conditions, $D_j = b$

3.8 Equivalent Stiffness

A popular notion in spring systems, electrical networks, and non-homogeneous heat conduction is to replace two or more members with a single "equivalent" member. Such a member would have an equivalent stiffness. Examples of this concept are often presented for springs or resistors in series or parallel. The evaluation of an equivalent stiffness is rather simple, but really loses its usefulness when an alternative solution like finite elements is available. However, the consideration of this concept can provide a useful review of the algebra of partitioned matrices.

As an example, return to the two springs shown in Fig. 3.5. The assembled equations given in Eq. (3.11) will be reordered to place the internal displacement, u_2, last in the equations. Then the matrix is partitioned as shown below:

$$\begin{bmatrix} k_a & 0 & -k_a \\ 0 & +k_b & -k_b \\ k_a & -k_b & (k_a+k_b) \end{bmatrix} \begin{Bmatrix} u_1 \\ u_3 \\ u_2 \end{Bmatrix} = \begin{Bmatrix} F_1 \\ F_3 \\ F_2 \end{Bmatrix} \qquad (3.38)$$

which could be denoted in symbolic form as in Eq. (3.14).

Now assume that the spring (or resistor) system has no externally applied force (or electrical potential) applied at the interior node. This is typical on the class of problem where hand solution methods utilize the concept of an equivalent spring (or resistor). Referring to Eq. (3.14) this means that $\mathbf{F}_g = \mathbf{0}$. However, to give our algebraic exercise more usefulness this condition will be postponed until later.

Recall the second row of the selected portions could be written as shown in Eq. (3.17). This would allow us to relate the "internal" variables, $\mathbf{u}_g$, to the "exterior" ones, $\mathbf{u}_r$. This is sometimes called a computational slave-master condition. From Eq. (3.17)

$$\mathbf{u}_g = \mathbf{K}_{gg}^{-1}(\mathbf{F}_g - \mathbf{K}_{gr}\mathbf{u}_r).$$

This master-slave relation could be substituted into the first row position of Eq. (3.15)

$$\mathbf{K}_{rr}\mathbf{u}_r = \mathbf{F}_r - \mathbf{K}_{rg}\mathbf{u}_g$$

$$= \mathbf{F}_r - \mathbf{K}_{rg}\mathbf{K}_{gg}(\mathbf{F}_g - K_{\mathbf{gr}}\mathbf{u}_r)$$

or combining the remaining terms involving $\mathbf{u}_r$

$$(\mathbf{K}_{rr} - \mathbf{K}_{rg}\mathbf{K}_{gg}^{-1}\mathbf{K}_{gr})\mathbf{u}_r = (\mathbf{F}_r - \mathbf{K}_{rg}\mathbf{K}_{gg}^{-1}\mathbf{F}_g)$$

or symbolically

$$\mathbf{K}^*_{rr}\mathbf{u}_r = \mathbf{F}^*_r$$

where $\mathbf{K}^*_{rr}$ and $\mathbf{F}^*_r$ are called condensed matrices. This process is usually referred to a **static condensation.** Its use is common in several more advanced finite element formulations.

Here, however, we are interested in an "equivalent" system when $\mathbf{F}_g = \mathbf{0}$ so

$$\mathbf{K}^*_{rr}\mathbf{u}_r = \mathbf{F}_r$$

and $\mathbf{K}^*$ is the desired equivalent stiffness (or resistance). For the previous two element examples the partitions yield:

$$K^* = \begin{bmatrix} k_a & 0 \\ 0 & k_b \end{bmatrix} - \begin{Bmatrix} -k_a \\ -k_b \end{Bmatrix} \frac{1}{k_a + k_b} \begin{bmatrix} -k_a & -k_b \end{bmatrix}$$

$$= \begin{bmatrix} k_a & 0 \\ 0 & k_b \end{bmatrix} - \begin{bmatrix} k_a^2 & k_a k_b \\ k_a k_b & k_b^2 \end{bmatrix} / (k_a + k_b)$$

$$K^*_{rr} = \frac{k_a k_b}{k_a + k_b} \begin{bmatrix} 1 & -1 \\ -1 & 1 \end{bmatrix}$$

Comparing with Eq. (3.4) verifies that $k_{equiv} = k_a k_b/(k_a + k_b)$ as expected.

3.9 Exercises

1. Noting that $d = (u_2 - u_1)$ verify, by expansion, that Eq. (3.1) is equivalent to the product $\mathbf{u}^T\mathbf{k}\mathbf{u}/2$. Refer to the notation of Eqs. (3.3) and (3.4).

2. Write a special matrix product program, say ATBA to evaluate the common expression $\mathbf{C} = \mathbf{A}^T\mathbf{B}\mathbf{A}$ where **B** is a square symmetric matrix and **A** is rectangular.
3. Solve the example in Fig. 3.6 when the force at point 2 acts to the left.
4. Write a program, say REQD, that will use Eq. (3.16) to return $\mathbf{u}_r$ when given $\mathbf{K}_{rr}$, $\mathbf{F}_r$, $\mathbf{K}_{rg}$, $\mathbf{u}_g$, and the sizes of $\mathbf{u}_r$ and $\mathbf{u}_g$, say NR and NG respectively. Utilize subroutines MATMLT, MATSUB, and SYMINV.
5. Write a program, say REACT, that will use Eq. (3.17) to return $\mathbf{F}_g$. Use MTMULT, MATMLT, and MATADD.
6. Consider the equations

$$\begin{bmatrix} 8 & 1 & 0 & 1 & 2 \\ 1 & 8 & 1 & 0 & 2 \\ 0 & 1 & 8 & 1 & 2 \\ 1 & 0 & 1 & 8 & 2 \\ 2 & 2 & 2 & 2 & 16 \end{bmatrix} \begin{Bmatrix} u_1 \\ u_2 \\ u_3 \\ u_4 \\ u_5 \end{Bmatrix} = \begin{Bmatrix} 24 \\ 37 \\ 40 \\ 46 \\ 110 \end{Bmatrix} .$$

Preform the static condensation to reduce these equations by eliminating the last degree of freedom, u_5.

MATHEMATICAL FORMULATIONS IN 1-D

4.1 Introduction

The previous chapter relied on physical intuition or basic conservation laws to formulate finite element models for several classes of problems. Some of the previous concepts can be carried over to two- or three-dimensional problems. However, it usually is less tedious to utilize a consistant mathematical concept to solve these more complicated problems. It was stated in the introductory chapter that the finite element method requires an integral formulation. In this chapter a few common methods that use integral procedures will be introduced.

4.2 A Sample ODE

We will introduce a sample ordinary differential equation, ODE, by means of a common physical problem. Consider a hot object with temperature T_0 that is surrounded by an environment at temperature T_r. Newton's law of cooling states that the time rate of cooling is proportional to the difference between the temperature of the object and the environment:

$$\frac{dT}{dt} = -k(T - T_r). \tag{4.1}$$

The initial condition is that $T(o) = T_0$. We desire the solution for time $t \geq 0$. The equation will be non-dimensionalized by using the change of variables that $y = (T -$

$T_r)/(T_0 - T_r)$ and $x = kt$. Then the ODE and initial conditions are

$$\frac{dy}{dx} + y = 0 \qquad \forall\, x \geqslant 0 \tag{4.2}$$

and $y = 1$ at $x = 0$. For future reference we will note that the exact solution of this problem is $y = e^{-x}$.

Generally we do not expect to get the exact solution to difficult engineering problems. However, we wish to get the best approximation possible. Thus, we will consider ways to approximate the solution to the above problem. Initially our approximate solution will be selected so as to satisfy the essential condition that $y = 1$ at $x = 0$. Thus assume that

$$y(x) = 1 + cx. \tag{4.3}$$

This describes the spatial variation of y and relates it to a single degree of freedom, c. The immediate task is to determine c. First we will restrict the region of interest to the range $0 \leqslant x \leqslant 1$. If we substitute the approximation into the ODE we define a **residual** error, R:

$$\frac{d}{dx}(1 + cx) + (1 + cx) = R \neq 0$$

or

$$R = 1 + c + cx.$$

The residual, R, will not vanish unless we happen to guess the exact solution. However, it is possible to compute a **least squares** estimate of the best value for c. That is accomplished by minimizing the integral of the square of R. That is,

$$I = \frac{1}{2}\int_0^1 R^2 dx \quad \text{is minimum.} \tag{4.4}$$

It was necessary to specify the domain of interest in order to establish the limits of integration.

To minimize the integral, I, with respect to the undetermined constant c we employ the usual calculus procedure of setting the derivative of I to zero and solve the resulting equation for c. Thus an alternative to Eq.(4.4) is the relation

$$\int_0^1 R\frac{\partial R}{\partial c}dx = 0. \tag{4.5}$$

Returning to Eq. (4.4), the present example gives

$$I = \frac{1}{2} \int_0^1 [(1 + c)^2 + 2(1 + c)cx + c^2x^2]dx$$

or

$$I = (1 + 3c + 7c^2/3)/2.$$

Minimizing with respect to c requires

$$\frac{\partial I}{\partial c} = 0 = (3 + 14c/3)$$

so that c = -9/14 and the approximate solution, on the domain $0 \leq x \leq 1$, is

$$y = 1 - 9x/14.$$

This approximation is compared with the exact solution in Fig. 4.1. The result is not very accurate. In current terminology this approach would be called a global approximation, or a single element solution. A logical question at this point is how to improve the accuracy of the approximate solution?

We will consider alternate mathematical formulations as well as the use of more undetermined constants. Note that Eq. (4.5) could be written as

$$\int_0^1 RWdx = 0 \tag{4.6}$$

where W is an error **weighting function.** In the previous least square procedure $W = \partial R/\partial c$. Clearly many other choices for W could be made. The most common alternate choices are weight functions that lead to **collocation methods, subdomain methods,** and the **Galerkin method.** The Galerkin method is now known to be extremely important in finite element analysis procedures. The weighting functions for these other methods are usually selected in the following ways:

(a) Collocation: Select as many points in the domain as there are undetermined dof. Then require the residual error to vanish at each of these points. In other words, the weight is the Dirac delta, $W = \delta(x_0)$. This procedure relies on the mathematical property of the Dirac delta that

$$\int f(x)\,\delta(x_0)dx = f(x_0). \tag{4.7}$$

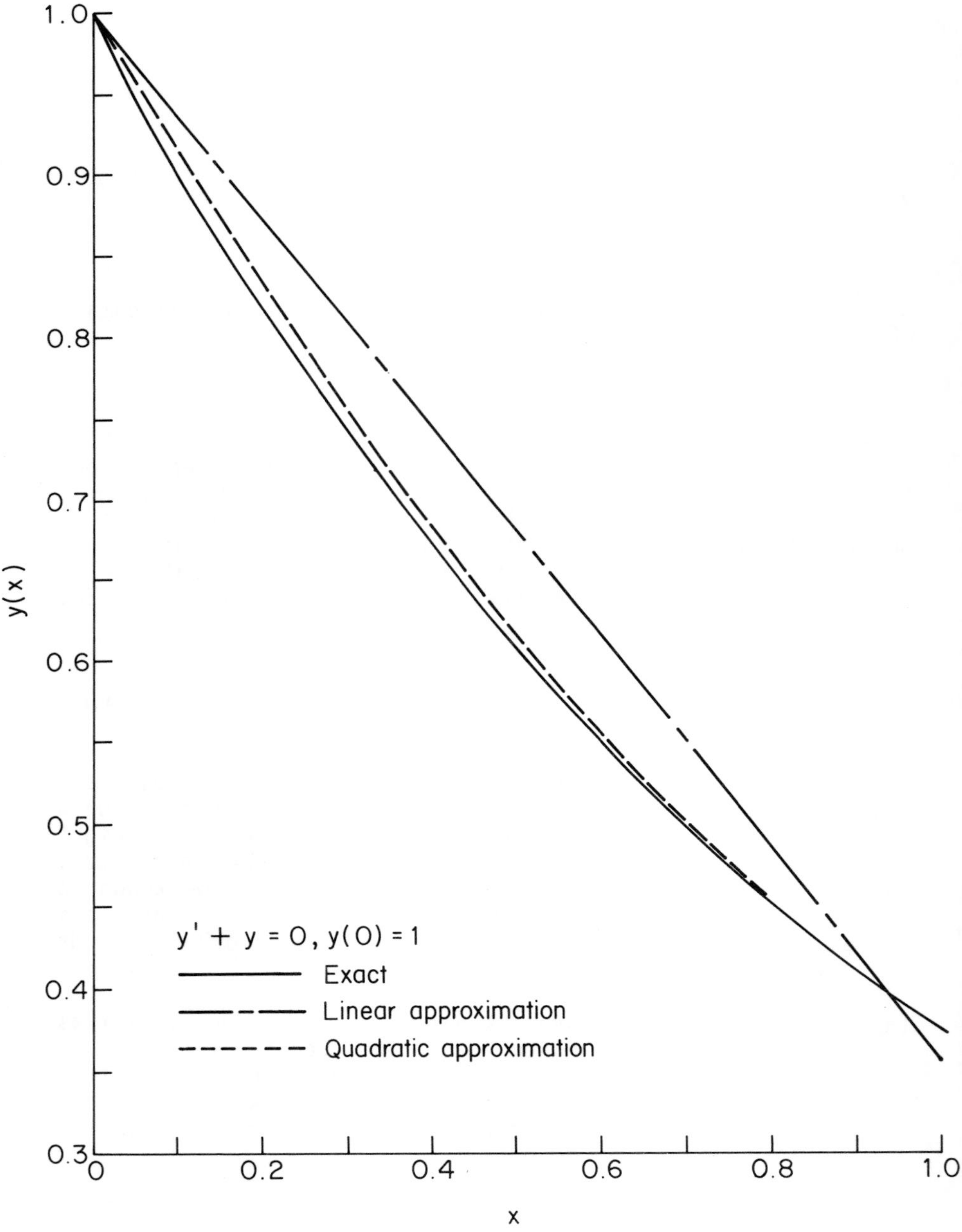

Fig. 4.1 Weighted Residual Global Approximations

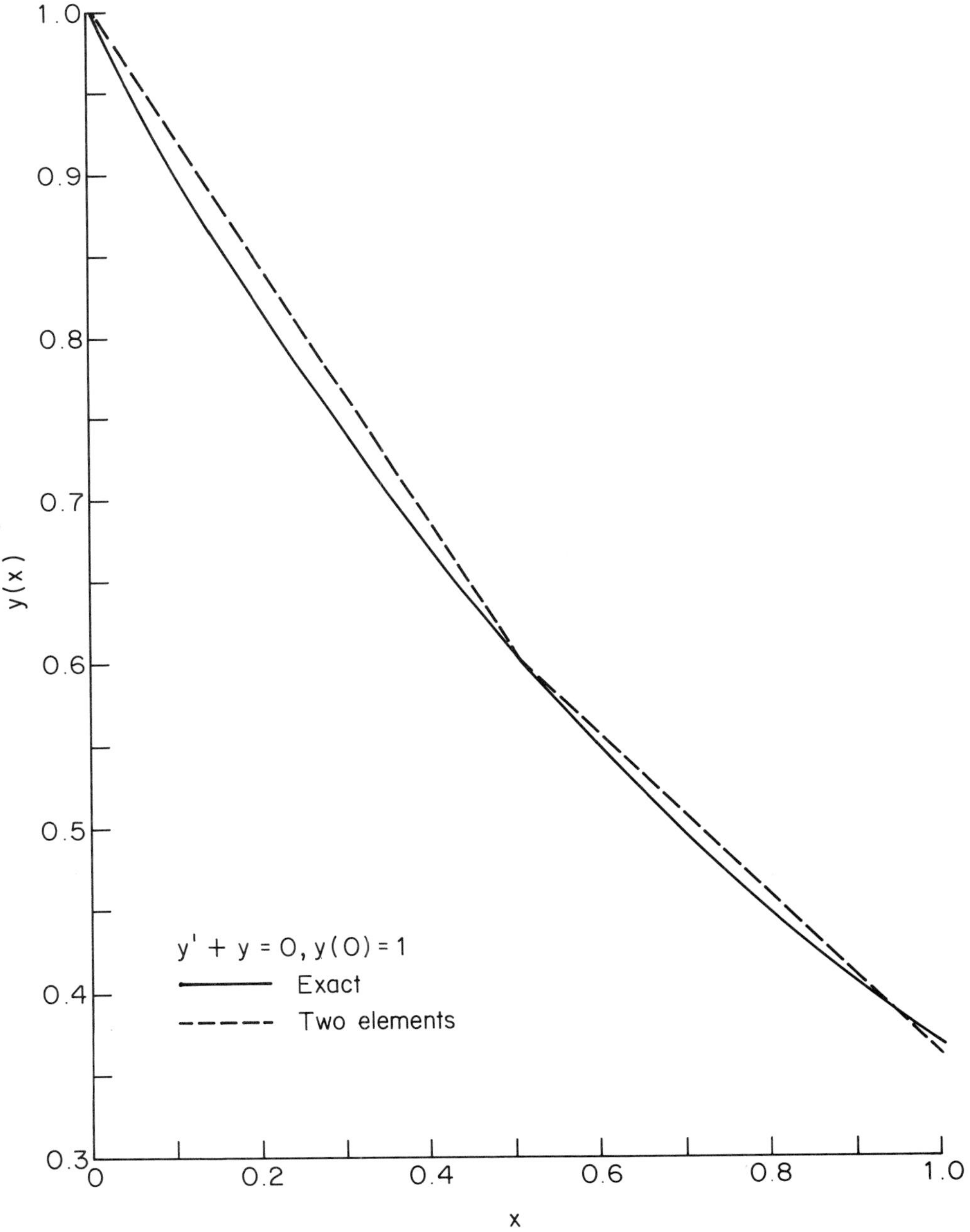

Fig. 4.2 A Piecewise Least Squares Approximation

(b) Subdomain: Divide the solution into as many subdomains as there are undetermined dof. Then require the integral of the residual to vanish on each subdomain. Thus the weight is unity in each subdomain and zero everywhere outside it.

(c) Galerkin: The weighting functions are taken to be the same spatial functions that were selected to multiply the undetermined dof in the approximate solution. A mathematical interpretation of this concept is that the weight function, W, is selected to be orthogonal to the residual error, R, over the solution domain.

Potential difficulties with both the collocation and subdomain methods are that they require the selection of arbitrary points or region locations within the solution domain. The accuracy of the results can vary significantly with one's experience with such procedures. Also they generate an unsymmetric set of equations to be solved for the undetermined dofs.

To illustrate these weighted residual procedures we will repeat the previous example with more degrees of freedom. Specifically we will pick a global quadratic approximation instead of the previous linear one, i.e.:

$$y(x) = 1 + c_1 x + c_2 x^2 \tag{4.8}$$

so that the residual error is

$$R(x) = 1 + c_1(1 + x) + c_2(2x + x^2). \tag{4.9}$$

Note that Eq. (4.8) has been selected to satisfy the essential boundary condition that $y = 1$ at $x = 0$.

Estimates for the two undetermined constants can be determined by any of the above weighted residual procedures. For example, if the previous least squares procedure is repeated one obtains

$$\frac{\partial R}{\partial c_1} = (1 + x)$$

$$\frac{\partial R}{\partial c_2} = (2x + x^2)$$

and the required equations are

$$\int_0^1 R\frac{\partial R}{\partial c_1}\ dx = 0 = 3/2 + 7c_1/3 + 9c_2/4$$

$$\int_0^1 R\frac{\partial R}{\partial c_1}\ dx = 0 = 4/3 + 9c_1/4 + 38c_2/15.$$

In matrix form this gives

$$\begin{bmatrix} 7/3 & 9/4 \\ 9/4 & 38/15 \end{bmatrix} \begin{Bmatrix} c_1 \\ c_2 \end{Bmatrix} = \begin{Bmatrix} -3/2 \\ -4/3 \end{Bmatrix}.$$

This shows the symmetric nature of the square coefficient matrix. Solving yields $c_1 = -0.9427$ and $c_2 = 0.3110$ and the least squares approximation is

$$y = 1 - 0.9427x + 0.3110x^2. \tag{4.10}$$

To apply the subdomain method we arbitrarily select two equal subdomains since there are two undetermined constants. Then we require

$$\int_0^{1/2} Rdx = 0 = 1/2 + 5c_1/8 + 7c_2/24$$

and

$$\int_{1/2}^1 Rdx = 0 = 1/2 + 7c_1/8 + 25c_2/24.$$

The unsymmetric equations are

$$\begin{bmatrix} 5/8 & 7/24 \\ 7/8 & 25/24 \end{bmatrix} \begin{Bmatrix} c_1 \\ c_2 \end{Bmatrix} = \begin{Bmatrix} -1/2 \\ -1/2 \end{Bmatrix}$$

and the computed solution is

$$y = 1 - 0.9474x + 0.3158x^2. \tag{4.11}$$

To formulate the Galerkin procedure rewrite Eq. (4.8) as

$$y = 1 + h_1(x)c_1 + h_2(x)c_2 \tag{4.12}$$

where $h_1 = x$ and $h_2 = x^2$ and require that

$$\int_0^1 Rh_1dx = 0 = 1/2 + 5c_1/6 + 11c_2/12$$

$$\int_0^1 Rh_2 dx = 0 = 1/3 + 7c_1/12 + 7c_2/10.$$

The resulting Galerkin approximation is

$$y = 1 - 0.9143x + 0.2857x^2. \tag{4.13}$$

Our final approximation will be generated by the collocation procedure. First it is necessary to select two points in the domain $0 \leq x \leq 1$. The easiest to pick are probably the third points, i.e., $x_1 = 1/3$ and $x_2 = 2/3$. However, from experience select two other symmetric points known as **Gauss points.** They are $x_1 = 0.211325$ and $x_2 = 0.788675$. Therefore, the necessary equations are

$$R(x_1) = 0 = 1 + 1.2113c_1 + 0.4673c_2$$

and

$$R(x_2) = 0 = 1 + 1.7887c_1 + 2.1994c_2$$

so that our final approximation is

$$y = 1 - 0.9474x + 0.3158x^2 \tag{4.14}$$

which happens to be the same as Eq. (4.11). Any other two points would give a different approximation. All of these quadratic approximations give less than two percent error at any point. The Galerkin solution is shown in Fig. 4.1 to illustrate a typical quadratic result.

Note that our original assumption in Eq. (4.8) was selected to automatically satisfy the forced boundary condition. It may not be practical to always do this. For example, we may wish to allow for any value to be assigned as the boundary condition at $x = 0$. To achieve this it is necessary to have more undetermined constants in our approximation. Thus an alternative to Eq. (4.8) as an approximate solution to Eq. (4.2) is to pick

$$y = a + c_1 x + c_2 x^2.$$

Then the model includes three undetermined constants, a, c_1, and c_2. Clearly three equations must be generated to evaluate these degrees of freedom. For simplicity select the collocation procedure and select the three points $x = 1/4$, $1/2$, and $3/4$. The resulting three equations are

$$R(1/4) = 0 = a + 5c_1/4 + 9c_2/16$$

$$R(1/2) = 0 = a + 3c_1/2 + 5c_2/4$$

$$R(3/4) = 0 = a + 7c_1/4 + 33c_2/16.$$

Expressing these in matrix form:

$$\frac{1}{16}\begin{bmatrix} 16 & 20 & 9 \\ 16 & 24 & 20 \\ 16 & 28 & 33 \end{bmatrix} \begin{Bmatrix} a \\ c_1 \\ c_2 \end{Bmatrix} = \begin{Bmatrix} 0 \\ 0 \\ 0 \end{Bmatrix}.$$

Now the desired boundary condition that a = 1 can be applied, via Eq. (3.16), to reduce the equations to

$$\frac{1}{16}\begin{bmatrix} 24 & 20 \\ 28 & 33 \end{bmatrix} \begin{Bmatrix} c_1 \\ c_2 \end{Bmatrix} = \begin{Bmatrix} -1 \\ -1 \end{Bmatrix}.$$

Then the result is $c_1 = -0.8966$ and $c_2 = 0.2759$. The main point of this example is that we can usually treat essential boundary conditions by introducing additional degrees of freedom.

These improved approximations are still basically global approximations and do not reflect the piecewise approximation nature of finite element solutions. To illustrate an alternate approach assume that the region of interest is replaced with an assemblage of finite elements. Let the length of a typical element be ℓ^e. Then we will replace the weighted residual in Eq. (4.6) with element contributions

$$\sum_e \int_{\ell^e} R^e W^e dx = 0 \tag{4.15}$$

where R^e is the residual error in the element and W^e is the weighting function. The summation sign implies a direct assembly process. Note that if there is more than one element then in general

$$\int_{\ell^e} R^e W^e dx \neq 0.$$

This means, for example, that one may minimize the weighted error over the total domain and yet not minimize it in every element. Or physically, one may have a system in external equilibrium and yet have some of its internal elements slightly out of equilibrium. These features occur because our solutions are approximate.

To establish R^e, or perhaps W^e, one must have the form of the approximate solution, say y. Usually this is generated by spatially interpolating the value of y, in element e, from its unknown nodal values, say $\mathbf{y}^e$. If we denote these interpolation functions (or shape functions, or basis functions) by the row matrix $\mathbf{H}^e(x)$ then the element approximation is

$$y^e(x) = H^e(x)\mathbf{y}^e \quad x \in \ell^e \tag{4.16}$$

where ℓ^e denotes the domain or region of the element and ε is read as "contained in." Substituting that approximation into the governing differential equation would define the residual error, R^e, in a typical element. It would also allow one to determine the weight functions, W^e, based on the least squares procedure. To obtain the weight functions for the Galerkin procedure one simply utilizes each of the terms in $\mathbf{H}^e(x)$. Of course, W^e is rather arbitrary in the case of collocation or subdomains. Thus we see it is the combination of Eqs. (4.15) and (4.16) that govern many finite element formulations.

To illustrate the piecewise approximation note that one could use two straight line segments rather than the single global quadratic expression utilized earlier. As an example of this concept assume that a typical element approximation varies linearly between the values, $y^e{}_1$, and $y^e{}_2$, at its nodes. Employing the notation of Eq. (4.16) this approximation could be written as an interpolation relation such as:

$$y^e(x) = y_1{}^e H_1{}^e(x) + y_2{}^e H_2{}^e(x) = \mathbf{H}^e\mathbf{y}^e, \quad x \varepsilon \ell^e \tag{4.17}$$

so that

$$dy^e/dx = y^e{}_1 dH^e{}_1/dx + y^e{}_2 dH^e{}_2/dx.$$

For example, if x_1 and x_2 denote the given coordinates of the element's nodes than these interpolation or shape functions can be written as

$$H_1(x) = \frac{x_2 - x}{x_2 - x_1}, \qquad H_2(x) = \frac{x - x_1}{x_2 - x_1} \tag{4.18}$$

where $\ell^e = (x_2 - x_1)$. It would also be possible to employ a non-dimensional element coordinate, say $s = (x - x_1)/\ell^e$, that is zero at node 1 and unity at node 2. In that case

$$H_1(s) = 1 - s, \quad H_2(s) = s$$

but that would require extra calculus since dy/dx = dy/ds*ds/dx, etc. Thus this approach will be delayed for a while.

To help in automating the weighted residual solutions we should have noted by now that Eq. (4.15) needs to be expanded to vector form. That is, to a set of simultaneous equations of the form

$$\sum_e \int_{\ell^e} R^e \mathbf{W}^e dx = \mathbf{0} \tag{4.19}$$

where for a least squares procedure

$$\mathbf{W}^e = \frac{\partial R^e}{\partial \mathbf{y}^e}$$

and for the Galerkin procedure

$$\mathbf{W}^e = \mathbf{H}^{e^T}.$$

For the subdomain procedure $\mathbf{W} = \mathbf{I}$ and for a collocation method $\mathbf{W} = \boldsymbol{\delta}$. In every case R^e results from a differential operator acting on y^e, and thus H^e, and it can be written as

$$R^e = \mathbf{G}^e(x)\mathbf{y}^e.$$

Here G results from the differential operator acting on H. For example, the global, or single element solution, residual in Eq. (4.9) could be written as

$$R^e = [1 \ (1 + x) \ (2x + x^2)] \begin{Bmatrix} 1 \\ c_1 \\ c_2 \end{Bmatrix}$$

or

$$R^e = \mathbf{G}^e(x)\mathbf{y}^e.$$

Akin [1] has shown that using the least squares procedure on the ordinary differential equation y' + ay = b gives

$$\int_{\ell^e} R^e \mathbf{W}^e dx = \mathbf{S}^e \mathbf{y}^e - \mathbf{C}^e$$

where the element square and column matrices are

$$\mathbf{S}^e = \frac{1}{3\ell}\begin{bmatrix} (3 - 3a\ell + a^2\ell^2) & (a^2\ell^2 - 6)/2 \\ \text{sym.} & (3 + 3a\ell + a^2\ell^2) \end{bmatrix} \tag{4.20}$$

and

$$\mathbf{C}^e = \frac{b}{2}\begin{Bmatrix} a\ell - 2 \\ a\ell + 2 \end{Bmatrix} \tag{4.21}$$

where ℓ denotes the element length. If we compare the above equation with our model we note that a = 1 and b = 0. Indeed Eq. (4.3) could be considered as a single element solution with $\ell = 1$ and with the boundary condition that $y_1 = 1$.

However, a typical finite element solution would use numerous elements. Figure 4.2 shows the exact solution and the results for a two element ($\ell = 1/2$) solution using Eq. (4.19). The two element solution has the same number of unconstrained dof's as did the previous global quadratic solution presented earlier. Thus they could be compared to illustrate the relative merits of the piecewise and global approximations.

The two element solution is obtained by assemblying Eq. (4.19) to yield

$$1/3\begin{bmatrix} 3.5 & -5.75 & 0 \\ -5.75 & 9.5+3.5 & -5.75 \\ 0 & -5.75 & 9.5 \end{bmatrix}\begin{Bmatrix} y_1 \\ y_2 \\ y_3 \end{Bmatrix} = \begin{Bmatrix} 0 \\ 0 \\ 0 \end{Bmatrix}$$

and then applying the condition that $y(0) = y_1 = 1$ to compute $y_2 = 0.6040$ and $y_3 = 0.3656$ which compare well with the exact values of 0.6065 and 0.3679, respectively.

4.3 A Second Order System

It is quite common to encounter engineering problems that are governed by second and fourth order differential equations. These are well suited to finite element solutions by weighted residuals and variational methods. Before considering variational methods we will extend some of the previous weighted residual methods to a typical second order system. As our model equation consider

$$k\frac{d^2y}{dx^2} + Q = 0, \quad 0 \leqslant x \leqslant L \tag{4.22}$$

where k could be considered to be a typical physical constant and Q is a forcing or source term. Both of these terms could be dependent on x. The system would also have to include two boundary conditions. The most common type of boundary condition is the essential, or forced, condition that y has a given value at a specified value of x. It is also possible to have natural boundary conditions which will be considered later.

To begin the construction of a finite element model we first select a typical element and its corresponding element dof's $\mathbf{y}^e$. Within this element the value of y^e is approximated, as in Eq. (4.16), as

$$y^e(x) = \mathbf{H}^e(x)\mathbf{y}^e = \mathbf{y}^{e^T}\mathbf{H}^{e^T}.$$

Recalling that

$$\frac{dy^e}{dx} = y^{e'} = \mathbf{H}^{e'}(x)\mathbf{y}^e$$

and

$$\frac{d^2y^e}{dx^2} = y^{e''} = \mathbf{H}^{e''}(x)\mathbf{y}^e = \mathbf{y}^{e^T}\mathbf{H}^{e''^T} \tag{4.23}$$

then the residual error in the typical element is

$$R^e(x) = k^e(x)\mathbf{H}^{e''}(x)\mathbf{y}^e + Q^e(x) \tag{4.24}$$

where k^e and Q^e are the values, or approximations, of k and Q, respectively, in typical element e. Once again we need to evaluate Eq. (4.19):

$$\sum_e \int_{\ell^e} R^e\mathbf{W}^e dx = 0$$

but we begin to encounter some limitations from calculus. To arrive at the above equation it was necessary to assume that

$$\int_\ell (*)dx = \sum_e \int_{\ell^e} (*)dx. \tag{4.25}$$

That assumption is not valid unless the integrand (*) satisfies a continuity condition across the boundaries where the ℓ^e join together. The condition is that if the integrand contains derivatives of order n, ie, d^ny/dx^n, then the function and all derivatives of order (n - 1) must be continuous.

In our analysis the integrand contains the product of R^e and $\mathbf{W}^e$. From Eq. (4.24) we note that R^e contains second derivatives (n = 2). Thus we are faced with the potential problem of utilizing an approximate y^e that has both y^e and y'^e continuous across the element interfaces. Such a solution would be called a **conforming** approximation. They are difficult to obtain in two- and three-dimensional problems but can be easily generated in one-dimension.

Since both the least squares method and the Galerkin procedure lead to symmetric sets of equations they are most popular. We will consider those two approaches from the point of view of how they affect the above continuity requirement. If we try least squares then

$$\mathbf{W}^e = \frac{\partial R^e}{\partial \mathbf{y}^e}$$

and it will have the same order derivatives as R^e. Thus the least squares method requires both y^e and $y^{e'}$ to be continuous across boundaries. That is known as C^1 continuity.

Next consider the Galerkin procedure where

$$\mathbf{W}^e = \mathbf{H}^{e^T}.$$

Since $\mathbf{H}^e$ is directly related to y^e it involves no derivatives so n = 0. Therefore, the product of $R^e\mathbf{W}^e$ involves a function with second derivatives and a function with no derivatives. If this product could be modified to reduce the order of the highest derivative then the above continuity requirement could be reduced. That would be very desirable.

Since calculus brought the potential difficulty to our attention it may also provide a way around the difficulty. This is indeed true for the Galerkin method. Recall the principal of **integration by parts**:

$$\int_a^b u\,dv = uv \Big|_a^b - \int_a^b v\,du \tag{4.26}$$

$$= u(x{=}b)\; v(x{=}b) - u(x{=}a)\; v(x{=}a) - \int_a^b v\,du.$$

Consider the integral

$$\int_a^b RW\,dx = 0$$

and let u = W, with a zeroth order derivative, and dv = Rdx where R has a second order derivative. Then

$$0 = uv \Big|_a^b - \int_a^b vdu \qquad (4.27)$$

and if the Galerkin method is utilized the integral has an integrand containing only derivatives of first order (n = 1). In that case the continuity requirement from calculus is that if we replace

$$\int_a^b vdu = \sum_e \int_{\ell^e} v^e du^e$$

then only the approximate function needs to be continuous across the element boundaries. That is called C^0 continuity. As will be discussed later the first term in Eq. (4.27) will help define any natural boundary conditions, at end points a and b, that the finite element solution will automatically satisfy.

To illustrate the use of integration by parts along with Galerkin's procedure return to Eq. (4.22) and employ a global weighting approximation. That is

$$\int_0^L (k\frac{d^2y}{dx^2} + Q)Wdx = 0$$

Consider the first integral and let u = W and dv = $ky''dx$. Integration by parts gives:

$$\int_0^L k\frac{d^2y}{dx^2} Wdx = kW\frac{dy}{dx} \Big|_0^L - \int_0^L kW \frac{dy}{dx}\frac{dW}{dx} dx$$

so that

$$\int_0^L (k\frac{dW}{dx}\frac{dy}{dx} - QW)dx - k\frac{dy}{dx} \Big|_0^L = 0. \qquad (4.28)$$

Before introducing a finite element model we can assign a physical meaning to this by noting that Eq. (4.22) will correspond to steady state heat transfer if k is the thermal conductivity, Q is the internal rate of heat generation, and y is the temperature.

Introducing an assemblage of finite elements and employing the Galerkin procedure generalizes Eq. (4.28) to

$$\sum_{e=1}^{NE} \int_{\ell^e} (k^e \frac{d\mathbf{W}}{dx}\frac{dy^e}{dx} - \mathbf{W}^e Q^e)dx$$

$$- k(L)\mathbf{W}(L)\frac{dy(L)}{dx} + k(o)\mathbf{W}(o)\frac{dy(o)}{dx} = \mathbf{0} \tag{4.29}$$

or

$$\sum_{e=1}^{NE} \int_{\ell^e} (k^e \frac{d\mathbf{H}^{e^T}}{dx}(\frac{d\mathbf{H}^e}{dx}\mathbf{y}^e) - \mathbf{H}^{e^T} Q^e)dx - \mathbf{C}_n = \mathbf{0}$$

where $\mathbf{C}_n$ contains the end conditions that will be considered in more detail later. If we note that $\mathbf{y}^e$ is constant for element e then the above relation can be written as

$$\sum_{e=1}^{NE} \mathbf{S}^e \mathbf{y}^e - \sum_{e=1}^{NE} \mathbf{C}_q^e - \mathbf{C}_n = \mathbf{0} \tag{4.30}$$

where the element square matrix (conductivity matrix) is

$$\mathbf{S}^e = \int_{\ell^e} k^e \frac{d\mathbf{H}^{e^T}}{dx}\frac{d\mathbf{H}^e}{dx}dx \tag{4.31}$$

and the element column vector (source vector) is

$$\mathbf{C}_q^e = \int_{\ell^e} \mathbf{H}^{e^T} Q^e dx. \tag{4.32}$$

At this point one may note that when the element has internal sources or variable properties it is much easier to establish the element matrices from a mathematical formulation than from physical intuition. From Eq. (4.28) we note that the integrand has only first derivatives. Therefore, the continuity requirement from calculus is that our approximations for y must be continuous between elements. If we select the values of y at the nodal points as unknowns then this continuity condition is automatically satisfied. This is true since the two elements on either side of the node share the value of y at the node. This concept was illustrated in

Fig. 1.1 and shown with more detail in Fig. 4.3. There are a number of interpolation functions which can satisfy the continuity requirement. The simplest is the linear interpolation given in Fig. 4.3 and listed earlier in Eqs. (3.8) and (4.18). Thus we can write Eq. (4.18) as

$$\begin{aligned} y^e(x) &= \mathbf{H}^e(x)\mathbf{y}^e \\ &= [H_1^e(x) \quad H^e{}_2(x)] \begin{Bmatrix} y_1^e \\ y_2^e \end{Bmatrix} \\ &= \left(\frac{x_2-x}{\ell^e}\right) y_1^e + \left(\frac{x-x_1}{\ell^e}\right) y_2^e . \end{aligned}$$

The first derivative of the approximation is

$$\frac{dy^e}{dx} = \frac{d\mathbf{H}^e}{dx}\,\mathbf{y}^e$$

where

$$\frac{d\mathbf{H}^e}{dx} = [-1/\ell^e \quad 1/\ell^e]$$

is a constant. Thus for this element selection Eq. (4.31) can be simplified to

$$\mathbf{S}^e = \frac{1}{\ell^{e^2}} \begin{bmatrix} 1 & -1 \\ -1 & 1 \end{bmatrix} \int_{\ell^e} k^e dx$$

and if the element conductivity is also constant it reduces to

$$\mathbf{S}^e = \frac{k^e}{\ell^e} \begin{bmatrix} 1 & -1 \\ -1 & 1 \end{bmatrix}$$

which agrees with the square matrix in Eq. (3.23) that was developed by physical intuition. The element source vector was not considered earlier. Substituting into Eq. (4.32) gives

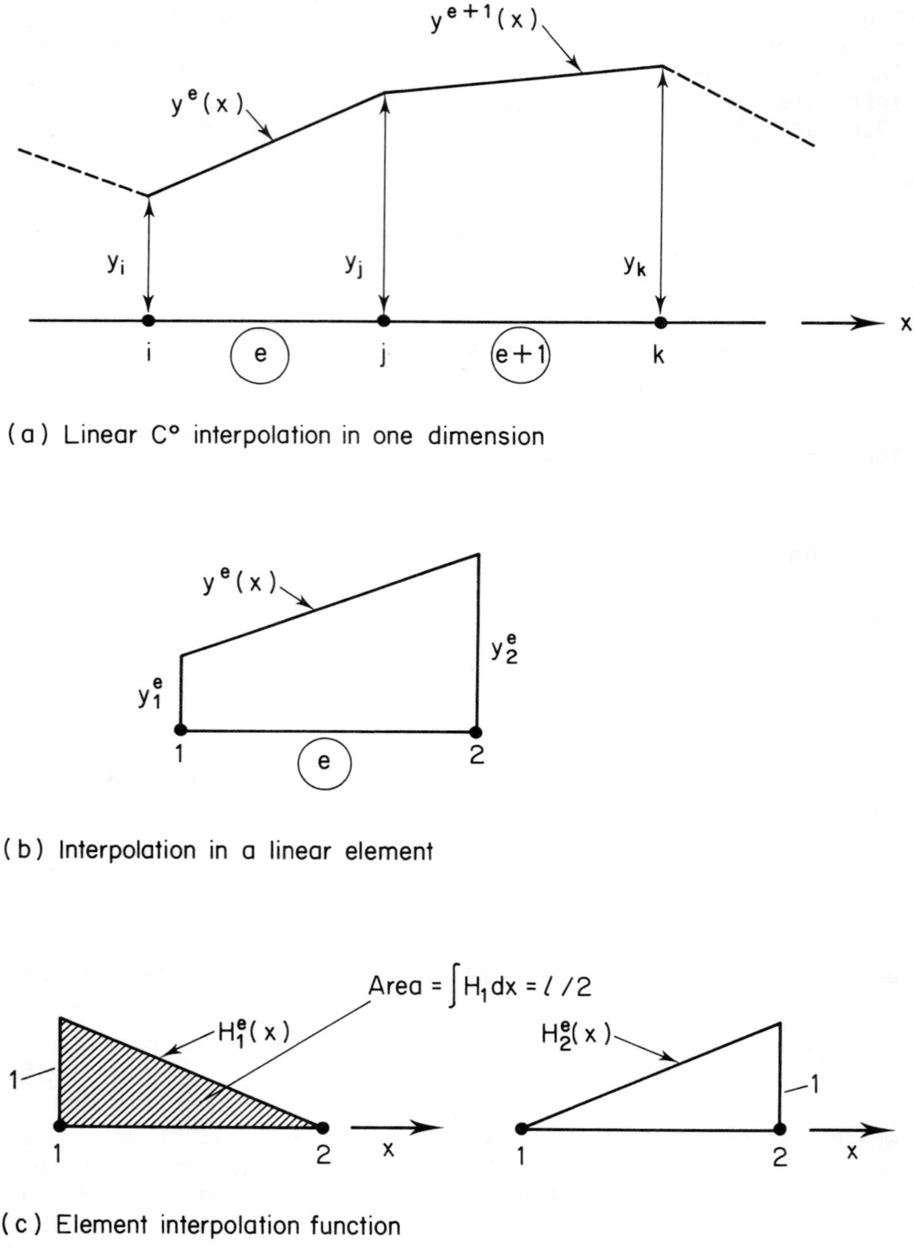

y^{e+1}(x)
y^e(x)
y_i
y_j
y_k
x
i
e
j
e+1
k
(a) Linear C° interpolation in one dimension
y^e(x)
y_2^e
y_1^e
1
e
2
(b) Interpolation in a linear element
Area = ∫H_1 dx = l/2
H_1^e(x)
1
1
2
x
H_2^e(x)
1
1
2
x
(c) Element interpolation function

Fig. 4.3 Piecewise linear interpolation

$$\mathbf{C}^e_q = \int_{\ell^e} \begin{Bmatrix} H_1^e(x) \\ H^e_2(x) \end{Bmatrix} Q^e dx.$$

Assuming a constant heat generation rate, Q, only $\mathbf{H}^e$ needs to be integrated to evaluate C^e. It may be useful to remember that the integral of a function equals the area under the curve. Thus from Fig. 4.3 we can write the triangular areas by inspection:

$$\int_{\ell^e} H^e_1 \, dx = \ell^e/2 = \int_{\ell^e} H^e_2 \, dx$$

and the source vector is

$$\mathbf{C}^e_q = \frac{q^e \ell^e}{2} \begin{Bmatrix} 1 \\ 1 \end{Bmatrix}.$$

This result should agree with physical intuition since $Q^e \ell^e$ is the total heat generated in the element and the above operation lumps half of this heat input to each end of the element.

Since both S^e and C^e are known we can consider the result of completing the summations in Eq. (4.30). The resulting equation could be written symbolically as

$$\mathbf{Sy} - \mathbf{C}_q - \mathbf{C}_n = \mathbf{0} \tag{4.33}$$

or simply

$$\mathbf{Sy} = \mathbf{C}$$

where **C** is the resultant source or forcing vector, **y** is the vector of all unknown nodal parameters, and **S** is the system square matrix. Equation (4.30) clearly shows that **C** and **S** are obtained by summing the contributions from all the $\mathbf{C}^e$ and $\mathbf{S}^e$ matrices. Since the $\mathbf{y}^e$ are subsets of **y** the direct assembly procedure discussed previously is used to carry out the assembly of these system matrices.

From Eq. (4.31) it is important to note that $\mathbf{S}^e$ and thus **S** are symmetric matrices. That characteristic would allow one to reduce the storage requirements for **S** and/or utilize more efficient algorithms to invert **S** or solve the governing system equations. However, at this point we are not interested in those computational aspects.

As an example of this Galerkin formulation, consider a rod that has its temperature fixed to zero at each end and has a uniform rate of heat generation along ts length. The system is shown in Fig. 4.4. However, due to symmetry it is only necessary to consider half the system as shown in Fig. 4.4b. At the plane of symmetry the temperature gradient, dy/dx, vanishes. That is, there is no conduction across the symmetry plane. Select a three element mesh. Then the boundary conditions are that y = 0 and dy/dx = 0 at node 1. The first is called a Dirichlett or essential condition and the second is a common Neumann condition that will be seen to be a natural boundary condition. Assume that the elements have equal lengths, and constant properties such that $\ell^e = 1m$, $k^e = 4$ w/mk, and $Q^e = 3$ w/m. Thus $\mathbf{S}^e$ and $\mathbf{C}^e$ are defined. However, returning to Eqs. (4.29) or (4.30) it is still necessary to evaluate $\mathbf{C}_n$. This involves the product of -kWdy/dx at the ends of the domain. Note that kdy/dx has an interpretation of a flux, say Q. In this case it is the heat flow at a point. Element one connects to the point x = 0 with node point 1. Therefore, at that point $W = H_1(0) = 1$. Similarly at the other end $W = H_2(x = L) = 1$ also. Thus the non-zero values of $\mathbf{C}_n$ are the fluxes, Q, at the ends of the domain. Therefore, $\mathbf{C}_n^T = [q_1 \quad 0 \quad 0 \quad -q_4]$. For these classes of problems one can either specify the function, y, is given such as at node 4 the flux, q_4, could be considered a reaction. At the symmetry plane the gradient and flux, q_1, vanishes. Therefore, $q_1 = 0$ and the condition of symmetry is naturally satisfied without additional calculations. This feature is a common advantage of finite element methods over other procedures such as finite difference methods. Carrying out the assembly gives

$$\begin{bmatrix} 4 & -4 & 0 & 0 \\ -4 & 8 & -4 & 0 \\ 0 & -4 & 8 & -4 \\ 0 & 0 & -4 & 4 \end{bmatrix} \begin{Bmatrix} y_1 \\ y_2 \\ y_3 \\ y_4 \end{Bmatrix} = \begin{Bmatrix} 1.5 \\ 3 \\ 3 \\ 1.5 \end{Bmatrix} + \begin{Bmatrix} -0 \\ 0 \\ 0 \\ q_4 \end{Bmatrix}$$

applying the essential condition that $y_4 = 0$ gives the reduced set

$$\begin{bmatrix} 4 & -4 & 0 \\ -4 & 8 & -4 \\ 0 & -4 & 8 \end{bmatrix} \begin{Bmatrix} y_1 \\ y_2 \\ y_3 \end{Bmatrix} = \begin{Bmatrix} 1.5 \\ 3 \\ 3 \end{Bmatrix}$$

so $\mathbf{y}^T = [3.375 \quad 3.0 \quad 1.875 \quad 0.0]$K. Determining the thermal reaction yields

$$-4y_3 + 4y_4 = 1.5 - q_4$$

$$-4(1.875) + 0 = 1.5 - q_4$$

$$q_4 = +9w$$

As expected this equals all of the internally generated heat, in half the structure, since it can only be conducted out the end point.

4.4 One Dimensional Variational Principles

Variational principles are widely used to formulate various physical and mathematical problems. They became popular after extensive development by Leonard Euler (1707-1783). They are not commonly utilized in undergraduate engineering education. The mathematical approach in a variational principle is to render stationary an integral equation so as to obtain the solution of a corresponding differential equation. Usually the stationary value is a minimum. The integrand, or functional, is a function of the coordinates, field amplitudes and their derivatives. Minimizing the integral also leads to appropriate boundary conditions associated with the differential equation. Finding the stationary value of a functional is a generalization of the elementary theory of the calculus of maxima and minima.

We are not interested in considering the mathematical details of variational principles. Here we will simply tabulate some of the more common principles associated with stable equilibrium problems. The differential equation, known as the **Euler equation,** associated with the variational principle will also be tabulated.

In ordinary calculus we are generally interested in functions. In other applications one encounters functions of functions. These quantities are called **functionals.** A typical form of a functional

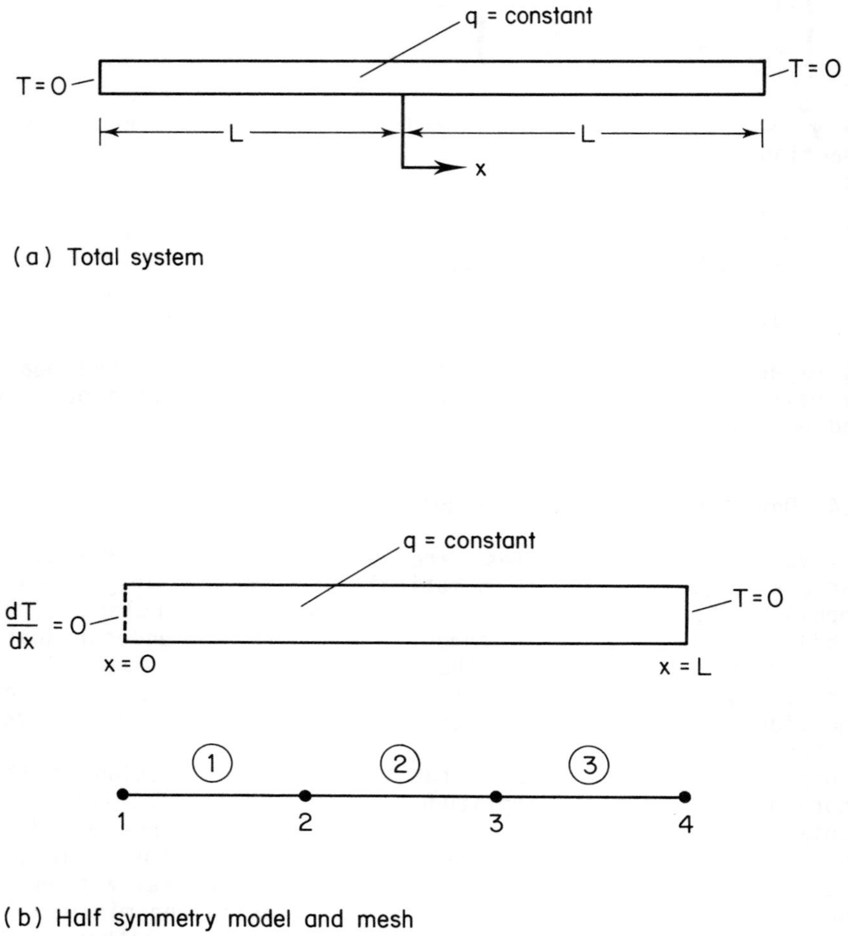

Fig. 4.4 A Steady State Conduction Problem

$$I = \int_0^L f(x, u, u_x)dx$$

where x is a spatial coordinate, u is a function of x, and $u_x = \partial u/\partial x$ is the partial derivative of u with respect to x.

Consider the problem of determining the minimum of the integral

$$I(y) = \int_a^b F(x, y(x), y'(x))dx \tag{4.34}$$

where the values of y(a) and y(b) are given and $y' = dy/dx$. The necessary condition for the minimum to exist is that y(x) satisfies the Euler equation

$$\frac{\partial F}{\partial y} - \frac{d}{dx}\left(\frac{\partial F}{\partial y'}\right) = 0 . \tag{4.35}$$

If the essential conditions y = y(a) or y = y(b) are not applied then the minimization gives the natural boundary conditions that

$$\frac{\partial F}{\partial y'} = 0 \tag{4.36}$$

at those points. As a specific example if

$$I(y) = 1/2 \int_0^1 (-y'^2 + y^2 + 2yx)dx$$

is minimized subject to the condition that y(0) = y(1) = 0 then

$$\frac{\partial F}{\partial y} = 2y + 2x$$

and

$$\frac{\partial F}{\partial y'} = -2y'$$

so the Euler equation is

$$(2y + 2x) - \frac{d}{dx}(-2y') = 0$$

or

$$y'' + y + x = 0.$$

The natural boundary condition would be that $y' = 0$ at $x = 0$ or $x = 1$. If the boundary condition at $x = 0$ was of the type $y'(x = 0) = g$ then Eq. (4.37) would be extended to include this condition by adding $gy(x = 0)$. The corresponding generalization of Eq. (4.34) would be

$$I = \int_a^b F(x,y,y')dx + g_b(x,y)\Big|_{x=b} - g_a(x,y)\Big|_{x=a} \tag{4.37}$$

and the natural boundary conditions for the Euler equation are

$$\frac{\partial F}{\partial y'} + \frac{\partial g_a}{\partial y} = 0 \qquad x=a$$

and

$$\frac{\partial F}{\partial y'} + \frac{\partial g_b}{\partial y} = 0 \qquad x=b. \tag{4.38}$$

We will also encounter variational principles such as

$$I(y) = \int_a^b F(x,y,y',y'')dx \tag{4.39}$$

where $y(a)$, $y'(a)$, $y(b)$, and $y'(b)$ are given. The corresponding Euler equation is

$$\frac{\partial F}{\partial y} - \frac{d}{dx}\left(\frac{\partial F}{\partial y'}\right) + \frac{d^2}{dx^2}\left(\frac{\partial F}{\partial y''}\right) = 0. \tag{4.40}$$

The associated natural boundary conditions are that

$$\frac{\partial F}{\partial y'} - \frac{d}{dx}\left(\frac{\partial F}{\partial y''}\right) = 0$$

if y is not specified, and

$$\frac{\partial F}{\partial y''} = 0$$

if y' is not specified. As a common example of the above functional consider

$$I(y) = \int_0^L (EI(y'')^2 - 2yq)dx \tag{4.41}$$

where y(0) = 0 and y(1) = 0. Substituting into Eq. (4.40)

$$\frac{\partial F}{\partial y} = -2q$$

$$\frac{\partial F}{\partial y'} = 0$$

$$\frac{\partial F}{\partial y''} = 2EIy''$$

so the governing Euler equation is

$$(-2q) - 0 + \frac{d^2}{dx^2}(2EIy'') = 0$$

or

$$EIy'''' = q \tag{4.42}$$

and since the essential conditions on y' were not applied the natural boundary conditions are y"(0) = 0 = y"(L). Thus the fourth order equation in Eq. (4.42) has the four boundary conditions it would require.

The reader probably has recognized Eq. (4.42) from mechanics of materials. It is the differential equation for the deflection, y, of a linear beam subjected to a transverse distributed load, q. In mechanics of materials we associate the following physical quantities with the mathematics:

y	deflection
y'	slope,
EIy"	moment, M
EIy"'	transverse shear, V = dM/dx
EIy""	load, q = dV/dx.

Therefore, in this example the natural boundary conditions require that the moments vanish at each end and the essential conditions require that the deflections vanish. Physically that corresponds to a simply supported beam. If a boundary condition had been specified on the slope, y', then the natural condition would not be in effect. That is, the moment (and curvature, y") does not vanish but becomes an unknown reaction when y' is given.

The previous examples are only a few of the many variational principles that are utilized in engineering and mathematics. Later we will consider the extension of variational principles to two- and three-dimensional problems. However, for the time being we are mainly interested in using variational procedures to assist in formulating our finite element models.

The variational principle provides the integral formulation required for a finite element analysis. We begin by introducing our finite element mesh and assuming

$$I = \sum_e I^e + \sum_b I^b \tag{4.43}$$

where I^e is a contribution from an integral over a typical element, ℓ^e, and I^b denotes any boundary contribution terms such as those added to Eq. (4.34) to form Eq. (4.37).

If we consider Eq. (4.34) the previous calculus continuity requirement then an approximate solution in I^e would need to be C^0. But Eq. (4.39) would require a C^1 approximation since it contains second derivatives.

In order to review the previous introductory concepts consider the following model problem which will serve as an analytic example. The differential equation of interest is

$$\frac{d^2t}{dx^2} + Q = 0 \tag{4.44}$$

on the domain, $0 < x < L$ and is subjected to the boundary conditions

$$t(L) = t_0 \quad \text{and} \quad \frac{dt}{dx}(0) = q. \tag{4.45}$$

It is closely related to the previous model equation given in Eq. (4.22). In this case the variational formulation states that the function, t, which satisfies the essential condition, $t(L) = t_0$, and minimizes

$$I = 1/2 \int_0^L ((dt/dx)^2 - 2tQ)dx + qt(0), \tag{4.46}$$

also satisfies Eqs. (4.44) and (4.45). Here I is assumed to be the sum of the NE element and NB boundary segment contributions so that

$$I = \sum_{e=1}^{NE} I^e + \sum_{b=1}^{NB} I^b.$$

where here NB = 1 and $I^b = qt(0)$ and a typical element contribution is

$$I^e = 1/2 \int_{\ell^e} ((dt^e/dx)^2 - 2Q^e t^e)dx,$$

where ℓ^e is the length of the element. To evaluate such a typical element contribution it is necessary to introduce the interpolation function, $\mathbf{H}^e$, such that once again

$$t^e(x) = \mathbf{H}^e(x)\mathbf{T}^e = \mathbf{T}^{e^T}\mathbf{H}^{e^T}$$

and

$$\frac{dt^e}{dx} = \frac{d\mathbf{H}^e}{dx}\mathbf{T}^e = \mathbf{T}^{e^T}\frac{d\mathbf{H}^{e^T}}{dx},$$

where $\mathbf{T}^e$ denotes the nodal values of t for element e. Thus a typical element contribution is

$$I^e = 1/2\, \mathbf{T}^{e^T}\mathbf{S}^e\mathbf{T}^e - \mathbf{T}^{e^T}\mathbf{C}^e, \tag{4.47}$$

where

$$\mathbf{S}^e = \int_{\ell^e} \frac{d\mathbf{H}^{e^T}}{dx} \frac{d\mathbf{H}^e}{dx} \, dx \tag{4.48}$$

and

$$\mathbf{C}^e = \int_{\ell^e} Q^e \mathbf{H}^{e^T} dx. \tag{4.49}$$

Note that the last two equations are basically identical to those obtained by the Galerkin procedure in Eqs. (4.31) and (4.32). If the domain, [0,L], contains nodes such that t(0) = T_1, then the non-essential (natural) boundary term is

$$\mathbf{I}^b = \mathbf{T}^{b^T} \mathbf{C}^b = T_1 q.$$

Clearly, both the element degrees of freedom, $\mathbf{T}^e$, and the boundary degrees of freedom, $\mathbf{T}^b$, are subsets of the total vector of unknown parameters, $\mathbf{T}$. That is, $\mathbf{T}^e \subset \mathbf{T}$ and $\mathbf{T}^b \subset \mathbf{T}$. Of course, the T^b are usually a subset of the T^e (i.e., $T^b \subset T^e$ and in higher dimensional problems $\mathbf{H}^b \subset \mathbf{H}^e$). The main point here is that I = I($\mathbf{T}$) and that fact must be considered in the summation in Eq. (4.43) and in the minimization. The consideration of the subset relations is merely a bookkeeping problem. This allows Eq. (4.43) to be written as

$$I = 1/2\ \mathbf{T}^T \mathbf{S} \mathbf{T} - \mathbf{T}^T \mathbf{C}, \tag{4.50}$$

where

$$\mathbf{S} = \sum_{e=1}^{NE} \mathbf{b}^{e^T} \mathbf{S}^e \mathbf{b}^e,$$

$$\mathbf{C} = \sum_{e=1}^{NE} \mathbf{b}^{e^T} \mathbf{C}^e + \sum_{b=1}^{NB} \mathbf{b}^{b^T} \mathbf{C}^b$$

and where $\mathbf{b}^e$ denotes a set of Boolean bookkeeping operations such that $\mathbf{u}^e = \mathbf{b}^e \mathbf{u}$. The combination of the summations and bookkeeping is commonly referred to as the assembly process. It is often simply (incorrectly) written as a summation. No uniformly accepted symbol has been developed to represent these operations. They were described earlier as the direct assembly procedure.

It is easily shown, from Eq. (2.18), that minimizing $I = I(\mathbf{T})$ leads to

$$\partial I/\partial \mathbf{T} = \mathbf{0} = \mathbf{ST} - \mathbf{C}, \tag{4.51}$$

as the governing algebraic equations to be solved for the unknown nodal parameters, $\mathbf{T}$. To be specific, consider a linear interpolation element with N = 2 nodes per element. If the element length is $\ell^e = x_2 - x_1$ then the element interpolation is given in Eq. (4.18). Therefore the element square matrix is simply

$$S^e = \frac{1}{\ell^e}\begin{bmatrix} 1 & -1 \\ -1 & 1 \end{bmatrix}.$$

Assuming that $Q = Q_0$, a constant, the column matrix is

$$C^e = \frac{Q_0 \ell^e}{2}\begin{Bmatrix} 1 \\ 1 \end{Bmatrix}.$$

The exact solution of the original problem for $Q = Q_0$ is

$$t(x) = t_0 + q(x-L) + Q_0(L^2-x^2)/2.$$

Since for $Q_0 \neq 0$ the exact value is quadratic and the selected element is linear, our finite element model can give only an approximate solution. However, for the homogeneous problem $Q_0 = 0$, the model can (and does) give an exact solution. To compare a finite element solution with the exact one, select a two element model. Let the elements be of equal length, $\ell^e = L/2$. Then the element matrices are the same for both elements. The assembly process yields, $\mathbf{ST} = \mathbf{C}$ as

$$\frac{2}{L}\begin{bmatrix} 1 & -1 & 0 \\ -1 & (1+1) & -1 \\ 0 & -1 & 1 \end{bmatrix}\begin{Bmatrix} T_1 \\ T_2 \\ T_3 \end{Bmatrix} = \frac{Q_0 L}{4}\begin{Bmatrix} 1 \\ (1+1) \\ 1 \end{Bmatrix} - \begin{Bmatrix} q \\ 0 \\ 0 \end{Bmatrix}.$$

However, these equations do not yet satisfy the essential boundary condition of $t(L) = T_3 = t_0$. After applying this condition the reduced equations are

$$\frac{2}{L}\begin{bmatrix} 1 & -1 \\ -1 & 2 \end{bmatrix}\begin{Bmatrix} T_1 \\ T_2 \end{Bmatrix} = \frac{Q_0 L}{4}\begin{Bmatrix} 1 \\ 2 \end{Bmatrix} - \begin{Bmatrix} q \\ 0 \end{Bmatrix} + \frac{2t_0}{L}\begin{Bmatrix} 0 \\ 1 \end{Bmatrix}.$$

The inverse of the reduced **S** is

$$\mathbf{S}^{-1} = \left(\frac{L^2}{4}\right)\frac{2}{L}\begin{bmatrix} 2 & 1 \\ 1 & 1 \end{bmatrix} .$$

Computing $\mathbf{S}^{-1}\mathbf{C}$ give the new right hand side of

$$\frac{L}{2}\left(\frac{Q_0 L}{4}\begin{Bmatrix} 2(1) + 1(2) \\ 1(1) + 1(2) \end{Bmatrix} - q\begin{Bmatrix} 2(1) + 0 \\ 1(1) + 0 \end{Bmatrix} + \frac{2t_0}{L}\begin{Bmatrix} 0 + 1(1) \\ 0 + 1(1) \end{Bmatrix}\right) .$$

Simplifying yields

$$\mathbf{T}_r = \begin{Bmatrix} T_1 \\ T_2 \end{Bmatrix} = \frac{Q_0 L^2}{8}\begin{Bmatrix} 4 \\ 3 \end{Bmatrix} - \frac{qL}{2}\begin{Bmatrix} 2 \\ 1 \end{Bmatrix} + t_0\begin{Bmatrix} 1 \\ 1 \end{Bmatrix} .$$

These are the exact nodal values as can be verified by evaluating the exact solution at $x = 0$ and $x = L/2$, respectively. Thus our finite element solution is giving an **interpolate** solution. That is, it interpolates the solution exactly at the node points and is approximate at all other points. For the homogeneous problem, $Q_0 = 0$, the finite element solution is exact at all points. These properties are common to other finite element problems. The exact and finite element solutions are illustrated in Fig. 4.5. Note from that figure that the derivatives are also exact at least at one point in each element. The **optimal** derivative sampling **points** are usually the Gauss points. For $Q = Q_0$ the center point derivatives are exact in this example.

These examples have shown that we will have numerous options for formulating finite element integral models. These procedures will require different types of interpolation functions. So far only the linear interpolation has been utilized. Clearly the generation of the element matrices will require some difficult integrations. The use of local coordinates and/or numerical integration will help simplify these procedures and go a long way toward actually automating them. However, we will delay these refinements for a while.

Next we will devote a fair amount of detail to the use of the variational principle of Minimum Total Potential Energy to reformulate the bar element. This example will illustrate the power of the more advanced procedures over that of physical intuition.

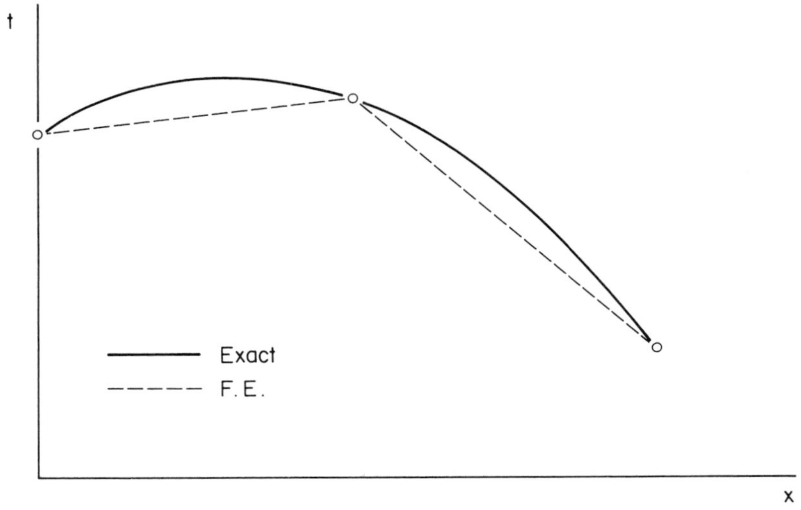

(a) Approximate and exact solutions

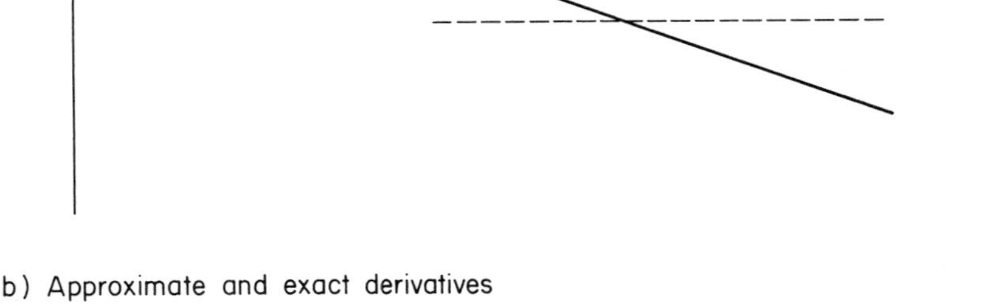

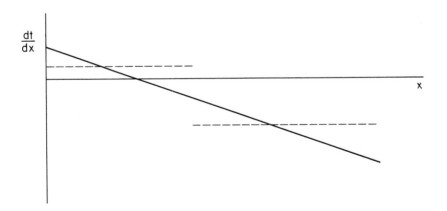

(b) Approximate and exact derivatives

Fig. 4.5 A Two Element Solution

4.5 Variational Formulations of the Bar Element

Most modern problems in elastic stress analysis are based on a variational principle. The principle of Minimum Total Potential Energy was briefly described in Sec. 3.3. To formulate the principle we will utilize the common displacement procedure. That is, the displacements of the structure are the primary unknowns. All other quantities of interest such as strains, stresses, or energy are to be related to the displacements.

Consider an axisymmetic rod shown in Fig. **4.6.** The cross-sectional area, $A(x)$, the perimeter, $p(x)$, the material modulus of elasticity, $E(x)$, and axial loading conditions would in general depend on the axial coordinate, x. The loading conditions could include surface tractions (shear) per unit area, $T(x)$, body forces per unit volume, $X(x)$, and concentrated point loads, P_i at point i.

The axial displacement at a point will be denoted by $u(x)$, and its value at point i is u_i. Recall that the work done by a force is the product of the force, the displacement at the point of application of the force, and the cosine of the angle between the force and the displacement. Here the forces are all parallel so the cosine is either plus or minus one. Evaluating the work

$$W = \int_0^L u(x)X(x)A(x)dx + \int_0^L u(x)T(x)p(x)dx + \sum_i u_i P_i . \qquad (4.52)$$

As mentioned earlier the total potential energy, π, includes the **strain energy**, U, and work of the externally applied forces, W. That is, $\pi = U - W$. In a mechanics of materials course it is shown that the strain energy per unit volume is half the product of the stress and strain. The axial strain and stress will be denoted by $\sigma(x)$ and $\varepsilon(x)$, respectively. Thus the strain energy is

$$U = 1/2 \int_0^L \sigma(x)\,\varepsilon(x)A(x)dx. \qquad (4.53)$$

The latter two equations have used $dV = Adx$ and $dS = pdx$ where dS is an exterior surface area. The work is clearly defined in terms of the displacement, u, since the loads would be given quantities. For example, the body force could be gravity, $X = \rho(x)g$, or a centrifugal load due to rotation about the y-axis, $X = \rho(x)x\omega^2$. Surface tractions are less

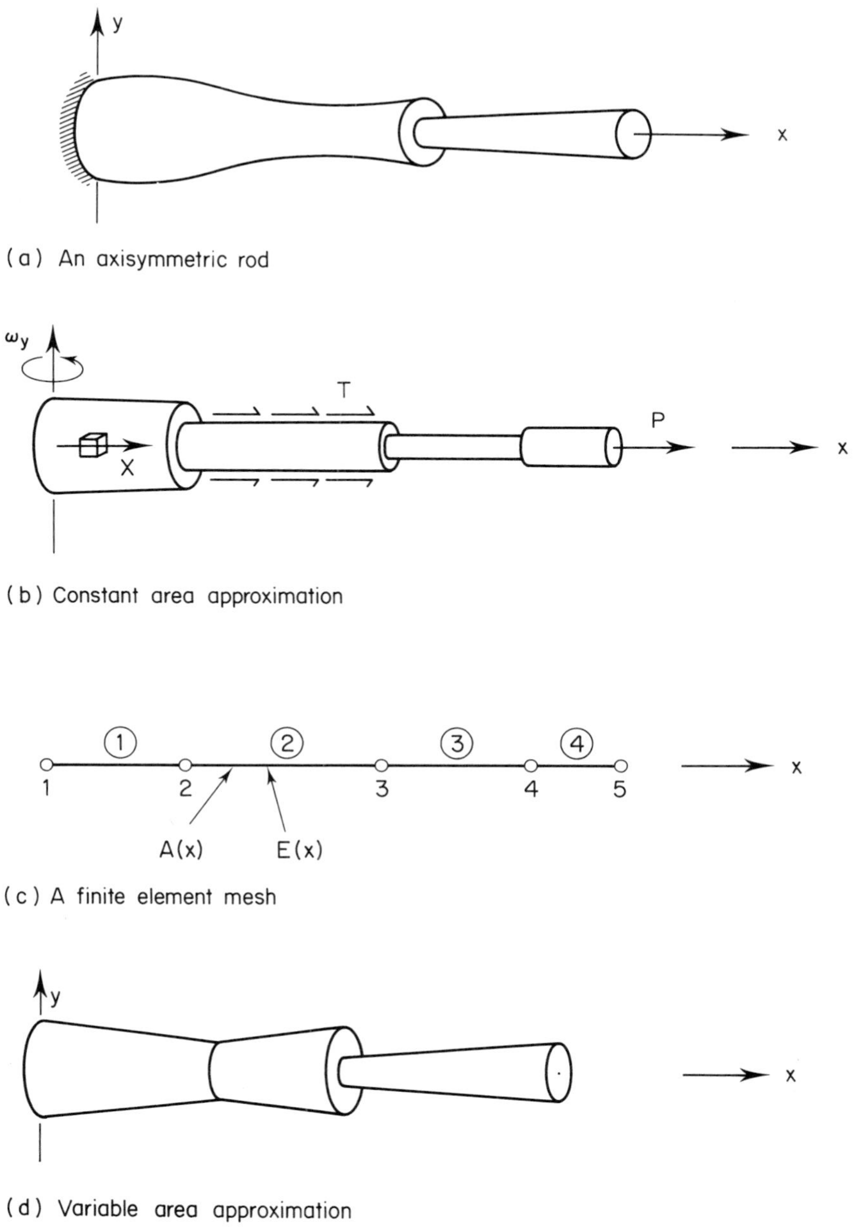

Fig. 4.6 A Typical Axially Loaded Bar

common in 1-D but it could be due to a very viscous fluid flowing over the outer surface and in the x-direction.

Our goal is to develop a displacement formulation. Thus we also need to relate both the stress and strain to the displacement, u. To begin with we recall the **strain-displacement relation**

$$\varepsilon(x) = \frac{du(x)}{dx} \tag{4.54}$$

which relates the strain to the derivative of the displacement. The stress at a point is directly proportional to the strain at the point. Thus it is also dependent on the **displacement gradient.** The relation between stress and strain is a **constitutive relation** known as **Hooke's law**

$$\sigma(x) = E(x)\varepsilon(x). \tag{4.55}$$

Therefore, we now see that the total potential energy depends on the unknown displacements and displacement gradients. We are searching for the displacement configuration that minimizes the total potential energy since that configuration corresponds to the state of stable equilibrium. As we have seen before, a finite element model can be introduced to approximate the displacements and their derivatives. Once again we begin by selecting the simplest model possible. That is, the two node line element with assumed linear displacement variation with x. As before we assume

$$\Pi = \sum_{e=1}^{NE} \Pi^e + \sum_{b=1}^{NB} \Pi^b + \sum_i P_i u_i \tag{4.56}$$

where Π^e is the typical element domain contribution, and Π^b is the contribution from a typical boundary segment domain. This one-dimensional example is somewhat misleading since in general a surface traction, T, acts only on a small portion of the exterior boundary. Thus the number of boundary domains, NB, is usually much less than the number of elements, NE. Here, however, NB = NE and the distinction between the two may not become clear until two-dimensional problems are considered.

Substituting Eqs. (4.52) and (4.53) into Eq. (3.5) and equating to Eq. (4.56) yields

$$\Pi^e = 1/2 \int_{\ell^e} \sigma^e \varepsilon^e A^e \, dx - \int_{\ell^e} u^e X^e A^e \, dx \tag{4.57}$$

$$\Pi^b = - \int_{\ell^e} u^{b^T} T^b p^b dx \tag{4.58}$$

where $u^e(x)$ and $u^b(x)$ denote the approximated displacements in the element and on the boundary surface, respectively. In this special example, $u^b = u^e$ but that is not usually true. Symbolically we interpolate such that

$$u^e(x) = H^e(x)u^e = \mathbf{u}^{e^T} \mathbf{H}^{e^T} \mathbf{(x)}$$

and

$$u^b(x) = H^b(x)\mathbf{u}^b = \mathbf{u}^{b^T} \mathbf{H}^{b^T} (\mathbf{x}).$$

Here we have the unusual case that $u^b = \mathbf{u}^e$ and $\mathbf{H}^b = \mathbf{H}^e$. Generally $\mathbf{u}^b$ is a subset of u^e (i.e., $\mathbf{u}^b \subset \mathbf{u}^e$) and $\mathbf{H}^b$ is a subset of $\mathbf{H}^e$. Substituting these interpolation relationships gives

$$\varepsilon^e(x) = \frac{du^e}{dx} = \frac{d\mathbf{H}^e(x)}{dx} \mathbf{u}^e = \mathbf{u}^{e^T} \frac{d\mathbf{H}^{e^T}(x)}{dx} \tag{4.59}$$

or in more common notation

$$\boldsymbol{\varepsilon}^e = \mathbf{B}^e(x)\mathbf{u}^e \tag{4.60}$$

$$\boldsymbol{\sigma}^e(x) = E^e(x)\ \boldsymbol{\varepsilon}^e(x) \tag{4.61}$$

$$\Pi^e = 1/2\ \mathbf{u}^{e^T} \int_{\ell^e} \mathbf{B}^{e^T} E^e \mathbf{B}^e A^e \mathbf{u}^e dx - \mathbf{u}^{e^T} \int_{\ell^e} \mathbf{H}^{e^T} X^e A^e dx$$

$$\Pi^b = -\mathbf{u}^{b^T} \int_{\ell^e} \mathbf{H}^{b^T} T^b p^b dx.$$

The latter two relations can be written symbolically as

$$\Pi^e = 1/2\ \mathbf{u}^{e^T} \mathbf{S}^e \mathbf{u}^e - \mathbf{u}^{e^T} \mathbf{C}_X^e \tag{4.62}$$

$$\Pi^b = -\ \mathbf{u}^{b^T} \mathbf{C}_T^b \tag{4.63}$$

where the element stiffness matrix is

$$\mathbf{S}^e = \int_{\ell^e} \mathbf{B}^{e^T} E^e \mathbf{B}^e A^e dx, \tag{4.64}$$

the element body force vector is

$$\mathbf{C}_X^e = \int_{\ell^e} \mathbf{H}^{e^T} X^e A^e dx, \tag{4.65}$$

and the boundary segment traction vector is

$$\mathbf{C}_T^b = \int_{\ell^e} \mathbf{H}^{b^T} T^b p^b dx. \tag{4.66}$$

The total potential energy of the system is

$$\Pi = 1/2 \sum_e \mathbf{u}^{e^T} \mathbf{S}^e \mathbf{u}^e - \sum_e \mathbf{u}^{e^T} \mathbf{C}_X^e - \sum_b \mathbf{u}^{b^T} \mathbf{C}_T^b - \mathbf{u}^T \mathbf{P} \tag{4.67}$$

where $\mathbf{u}$ is the vector of all of the unknown nodal displacements. Here we have assumed that the external point loads are applied at node points only. The last term represents the scalar, or dot, product of the nodal displacements and nodal forces. That is,

$$\mathbf{u}^T \mathbf{P} = \mathbf{P}^T \mathbf{u} = \sum_i u_i P_i.$$

Of course in practice most of the P_i are zero. By again applying the direct assembly procedure, or from the Boolean assembly operations of Eq. (4.50) the total potential energy is

$$\Pi(\mathbf{u}) = 1/2\ \mathbf{u}^T \mathbf{S} \mathbf{u} - \mathbf{u}^T \mathbf{C}$$

and minimizing with respect to all the unknown displacements, $\mathbf{u}$, gives the algebraic equilibrium equations for the entire structure

$$\mathbf{S}\mathbf{u} = \mathbf{C}.$$

Therefore, we see that our variational principle has lead to a very general and powerful formulation for this class of structures. It automatically includes features such as variable material properties, variable loads, etc. These were difficult to treat when relying solely on physical intuition. Although we will utilize the simple linear element none of our equations are restricted to that definition of **H**.

If we substitute $\mathbf{H}^e$ for the linear element and assume constant properties the element and boundary matrices are simple to evaluate. The results are

$$\mathbf{S}^e = \frac{E^e A^e}{\ell^e} \begin{bmatrix} 1 & -1 \\ -1 & 1 \end{bmatrix}$$

$$\mathbf{C}_X^e = \frac{X^e A^e \ell^e}{2} \begin{Bmatrix} 1 \\ 1 \end{Bmatrix}$$

$$\mathbf{C}_T^b = \frac{T^e P^e \ell^e}{2} \begin{Bmatrix} 1 \\ 1 \end{Bmatrix}.$$

Also in this case one obtains

$$\varepsilon^e = \frac{1}{\ell^e} [-1 \quad 1] \begin{Bmatrix} u_1^e \\ u_2^e \end{Bmatrix} \tag{4.68}$$

which means that the strain is constant in the element but the displacement approximation is linear. It is common to refer to this element as the constant strain, CS, element. The above stiffness matrix is the same as that obtained in Sec. 3.6.2. The load vectors take the resultant element, or boundary, force and place half at each node. That logical result does not carry over to more complicated load conditions and it then becomes necessary to rely on the mathematics of Eqs. (4.65) and (4.66).

As an example of a slightly more difficult loading condition consider a case where the body force varies linearly with x. This could include the case of centrifugal loading mentioned earlier. For simplicity assume a constant area A^e and let us define the value of the body force at each node of the element. To define the body force at any point in the element we again utilize the interpolation function and set

$$X^e(x) = \mathbf{H}^e(x)\mathbf{X}^e \tag{4.69}$$

where $\mathbf{X}^e$ are the defined nodal values of the body force. For these assumptions the body force vector becomes

$$\mathbf{C}_X^e = A^e \int_{\ell^e} \mathbf{H}^{e^T} \mathbf{H}^e dx \; \mathbf{X}^e .$$

For the linear element the integration reduces to

$$C_X^e = \frac{A^e \ell^e}{6} \begin{bmatrix} 2 & 1 \\ 1 & 2 \end{bmatrix} \begin{Bmatrix} X_1^e \\ X_2^e \end{Bmatrix} .$$

This agrees with our previous result for constant loads since if $X_1^e = X_2^e = X^e$ then

$$C_X^e = \frac{A^e \ell^e X^e}{6} \begin{Bmatrix} 2+1 \\ 1+2 \end{Bmatrix} = \frac{A^e \ell^e X^e}{2} \begin{Bmatrix} 1 \\ 1 \end{Bmatrix} .$$

A more common problem is the one illustrated in Fig. 4.6d where the area of the member varies along the length. To approximate that case, with constant properties, one could interpolate for the area at any point as

$$A^e(x) = \mathbf{H}^e(x)\mathbf{A}^e$$

then the stiffness in Eq. (4.64) becomes

$$\mathbf{S}^e = \frac{E^e}{(\ell^e)^2} \begin{bmatrix} 1 & -1 \\ -1 & 1 \end{bmatrix} V^e$$

where

$$V^e = \int_{\ell^e} A^e dx = \int_{\ell^e} \mathbf{H}^e(x) dx \; \mathbf{A}^e$$

$$= \frac{\ell^e}{2} [1 \quad 1] \begin{Bmatrix} A_1^e \\ A_2^e \end{Bmatrix} = \ell^e (A_1^e + A_2^e)/2$$

is the average volume of the element. In a similar manner the body force vector would be

$$\mathbf{C}_X^e = \frac{X^e}{2} \begin{Bmatrix} 1 \\ 1 \end{Bmatrix} V^e = \frac{X^e \ell^e (A_1^e + A_2^e)}{4} \begin{Bmatrix} 1 \\ 1 \end{Bmatrix} .$$

The above approximations should be reasonably accurate. However, we recall that the area is related to the radius by $A = \pi r^2$. Thus it would be slightly more accurate to describe the radius at each end and interpolate

$$r^e(x) = \mathbf{H}^e(x)\mathbf{r}^e = \mathbf{r}^{e^T}\mathbf{H}^{e^T}$$

so that

$$V^e = \int_{\ell^e} A^e dx = \pi \int_{\ell^e} r^e(x)^2 dx$$

$$V^e = \mathbf{r}^{e^T} \pi \int_{\ell^e} \mathbf{H}^{e^T}\mathbf{H}^e d\mathbf{x}\ \mathbf{r}^e$$

$$V^e = \mathbf{r}^{e^T} \frac{\pi \ell^e}{6} \begin{bmatrix} 2 & 1 \\ 1 & 2 \end{bmatrix} \mathbf{r}^e$$

or

$$V^e = \pi \ell^e (r_1^2 + r_1 r_2 + r_2^2)/3.$$

Clearly as one utilizes more advanced interpolation functions the integrals involved in Eqs. (4.64) to (4.66) become more difficult to evaluate. Before we show how to simplify these integrals we will consider an example problem to illustrate the use of these element matrices and to begin to introduce the benefits of post-solution calculations.

Consider a prismatic bar of steel rigidly fixed to a bar of brass and subjected to a vertical load of P = 10,000 lb. The structure is supported at the top point and is also subjected to a gravity (body force) load. Determine the deflections, reactions, and stresses for the properties tabulated in Fig. 4.7b. The first element has a stiffness constant of $EA/\ell = 0.7143 \times 10^6$ lb/in and the body force is $XA\ell = 1188.6$ lbs while for the second the corresponding terms are 0.4333×10^6 lb/in and 576. lbs respectively. The system nodal force vector is $\mathbf{P}^T = [R \quad 0 \quad 10{,}000]$ lb. Where R is the unknown reaction at node 1. Assemblying the equations gives

$$10^5\begin{bmatrix} 7.143 & -7.143 & 0 \\ -7.143 & 7.143 + 4.333 & -4.333 \\ 0 & -4.333 & 4.333 \end{bmatrix}\begin{Bmatrix} u_1 \\ u_2 \\ u_3 \end{Bmatrix}$$

$$= \begin{Bmatrix} R \\ 0 \\ 10{,}000 \end{Bmatrix} + 1/2 \begin{Bmatrix} 1188.6 \\ 1188.6 + 576. \\ 576. \end{Bmatrix} = \begin{Bmatrix} R + 594.3 \\ 882.3 \\ 10288 \end{Bmatrix}.$$

Applying the essential condition that $u_1 = 0$

$$10^5 \begin{bmatrix} 11.476 & -4.333 \\ -4.333 & 4.333 \end{bmatrix}\begin{Bmatrix} u_2 \\ u_3 \end{Bmatrix} = \begin{Bmatrix} 882.3 \\ 10288 \end{Bmatrix}$$

so $u_2 = 1.5638 \times 10^{-2}$ in., $u_3 = 3.9381 \times 10^{-2}$ in., and determining the reaction from the first system equation:

$$R = -11764.6 \text{ lb.}$$

This reaction is compared with the applied loads in Fig. 4.7c.

Now that all the displacements are known we can post process the results to determine the other quantities of interest. Substituting into the element strain-displacement relation, Eq. (4.68), gives

$$\varepsilon^1 = \frac{1}{420}\,[-1 \quad 1]\begin{Bmatrix} 0.0 \\ 0.01564 \end{Bmatrix} = 3.724 \times 10^{-5} \text{in/in.}$$

$$\varepsilon^2 = \frac{1}{240}\,[-1 \quad 1]\begin{Bmatrix} 0.01564 \\ 0.03938 \end{Bmatrix} = 9.892 \times 10^{-5} \text{in/in.}$$

and from Eq. (4.61) the element stresses are

$$\sigma^1 = E^1\varepsilon^1 = 30 \times 10^6(3.724 \times 10^{-5}) = 1117 \text{ lb/in}^2$$

$$\sigma^2 = E^2\varepsilon^2 = 13 \times 10^6(9.892 \times 10^{-5}) = 128.6 \text{ lb/in}^2.$$

These approximate stresses are compared with the exact stresses in Fig. 4.7d. This suggests that if accurate stresses are important then more elements are required to get good estimates from the piecewise constant element stress approximations. Note that the element stresses are exact if they are considered to act only at the element center.

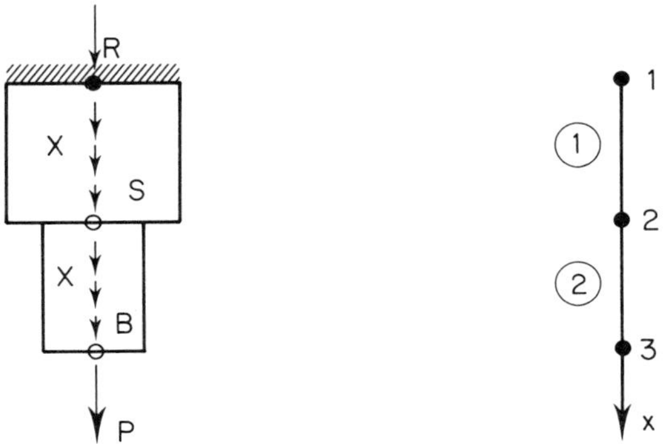

(a) Structure and a two element model

Element	ℓ^e	A^e	E^e	X^e
1	420"	10 sq. in.	30×10^6 psi	0.283 lb/in^3
2	240"	8 sq. in.	13×10^6 psi	0.300 lb/in^3

(b) System properties

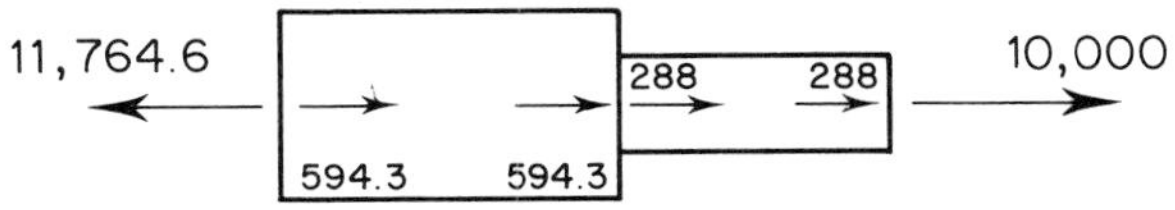

(c) Reaction and consistent loads

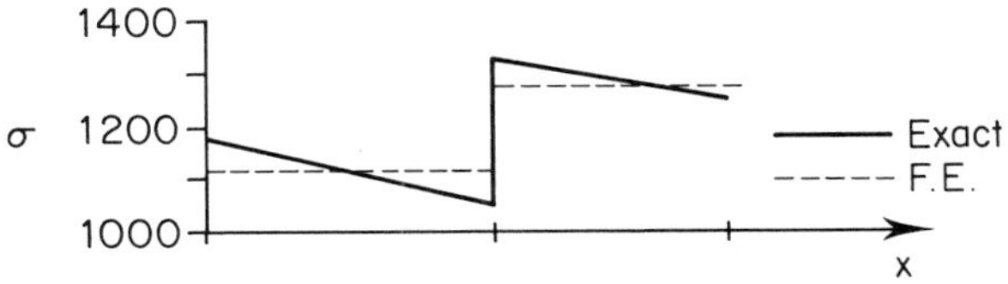

(d) Stress distribution

Fig. 4.7 An Axially Loaded System

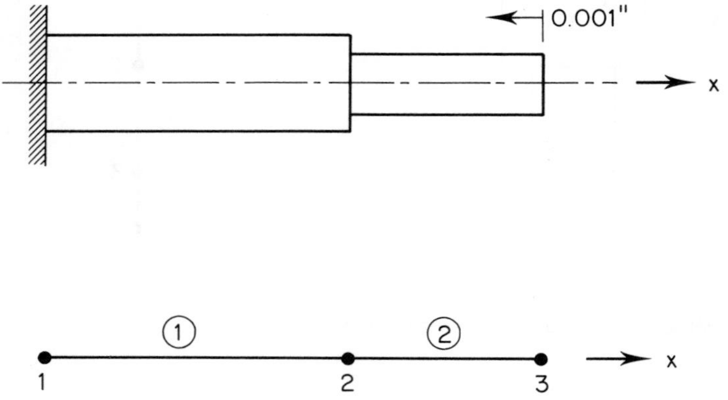

(a) Structure and mesh

Element	l^e	A^e	E^e	Δt^e	α^e
1	420"	10 sq. in.	30×10^6 psi	-35 F	6.7×10^{-6} in/inF
2	240"	8 sq. in.	13×10^6 psi	-35 F	12.5×10^{-6} in/inF

(b) System properties

Fig. 4.8 A Thermally Loaded Structure

Before leaving the bar element it may be useful to note that another common loading condition can be included. That is the loading due to an initial thermal strain. Recall that the thermal strain, ε_t, due to a temperature rise of Δt is $\varepsilon_t = \alpha\Delta t$ where α is the coefficient of thermal expansion. The work term in Eq. (4.52) is extended to include this effect by adding

$$W_t = \int_0^L \sigma\varepsilon_t A(x)dx.$$

This defines an element thermal force vector

$$\mathbf{C}_t^e = \int_{\ell^e} \mathbf{B}^{e^T}(x)E^e(x)\alpha^e(x)\Delta t(x)A^e(x)dx$$

or for constant properties and uniform temperature rise

$$\mathbf{C}_t^e = E^e\alpha^e \Delta t^e A^e \begin{Bmatrix} -1 \\ +1 \end{Bmatrix}. \qquad (4.70)$$

There is a corresponding change in Eq. (4.61) such that

$$\sigma = E(\varepsilon - \varepsilon_t) .$$

As a numerical example of this loading condition consider the statically indeterminate system shown in Fig. 4.7. The left support is fixed but the right support is displaced to the left 0.001 in. and the system is cooled by 35 F. Find the stress in each member. The assembled equations are

$$10^5 \begin{bmatrix} 7.143 & -7.143 & 0 \\ -7.143 & 11.476 & -4.333 \\ 0 & -4.333 & 4.333 \end{bmatrix} \begin{Bmatrix} u_1 \\ u_2 \\ u_3 \end{Bmatrix}$$

$$= 10^4 \begin{Bmatrix} -7.035 \\ 7.035 - 4.550 \\ +4.550 \end{Bmatrix} + \begin{Bmatrix} R_1 \\ 0 \\ R_2 \end{Bmatrix}$$

applying the boundary conditions that $u_1 = 0$, and $u_3 = -0.001$ in. and solving for u_2 yields $u_2 = -0.02203$ in. The reactions at points 1 and 3 are $R_1 = -54613$ lb, and $R_3 = +54613$ lb. Substituting into the element strain-displacement matrices yields element strains of -5.245×10^{-5} in/in, and $+8.763 \times 10^{-5}$ in/in, respectively.

4.6 Variational Formulation for 1-D Heat Transfer

A problem closely related to the previous problem is that of steady state heat transfer. Consider the heat transfer in a slender rod that has a specified temperature, t_0, at $x = 0$ and is insulated at the other end, $x = L$. The rod has cross-sectional area, A, with a thermal conductivity of K. Thus the rod conducts heat along its length. The rod is also surrounded by a convecting medium with a uniform temperature of t_∞. Thus the rod also convects heat on its outer surface area. Let the convective transfer coefficient be h and the outer perimeter of the rod be P. To simplify the model we will assume that the external reference temperature is $t_\infty = 0$. The governing differential equation for the temperature, $t(x)$, is given by Myers [16] as

$$KA\frac{d^2t}{dx^2} - hPt = 0, \quad 0 < x < L \tag{4.72}$$

with the essential condition

$$t(0) = t_0$$

and the natural boundary condition

$$\frac{dt}{dx}(L) = 0.$$

The exact solution can be shown to $t(x) = \cosh(m(L - x)) / \cosh(mL)$ where $m^2 = hP/KA$.

This problem can be identified as the Euler condition of a variational principle. This principle will lead to system equations that are structured differently from our previous example with the bar. In that case, the boundary integral contributions (tractions) defined a column matrix and thus went on the right hand side of the system equations. Here we will see that the boundary contributions (convection) will define a square matrix. Thus they will go into the system coefficients on the left hand side of the system equations.

Generally a variational formulation of steady state heat transfer involves volume integrals containing conduction terms and surface integrals with boundary heat flux, eq., convection, terms. In our one-dimensional example both the volume and surface definitions involve an integral along the length of the rod. Thus the distinction between volume and surface terms is less clear and the governing functional given by Myers is simply stated as a line integral. Specifically, one must render stationary the functional

$$I(t) = 1/2 \int_0^L (KA(\frac{dt}{dx})^2 + hPt^2)dx \qquad (4.73)$$

subject to the essential condition at x=0.

Divide the rod into a number of nodes and elements and introduce a finite element model where we assume

$$I = \sum_{e=1}^{NE} I^e + \sum_{b=1}^{NB} I^b$$

where we have defined a typical element volume contribution of

$$I^e = 1/2 \int_{\ell^e} K^e A^e (\frac{dt^e}{dx})^2 dx \qquad (4.74)$$

and typical boundary contribution is

$$I^b = 1/2 \int_{\ell^b} h^b p^b t^{b^2} dx. \qquad (4.75)$$

Using our interpolation relations as before

$$t^e(x) = \mathbf{H}^e(x)\mathbf{t}^e = \mathbf{t}^{e^T}\mathbf{H}^{e^T}$$

and again in this special case $t^b(x) = t^e(x)$. Thus these can be written symbolically as

$$I^e = 1/2\ \mathbf{t}^{e^T}\mathbf{S}^e\mathbf{t}^e$$

and

$$I^b = 1/2\ \mathbf{t}^{b^T}\mathbf{S}^b\mathbf{t}^b\ .$$

Assuming constant properties the square matrices reduce to

$$S^e = \frac{K^e A^e}{\ell^e} \begin{bmatrix} 1 & -1 \\ -1 & 1 \end{bmatrix}$$

and

$$S^b = \frac{h^b P^b \ell^b}{6} \begin{bmatrix} 2 & 1 \\ 1 & 2 \end{bmatrix} .$$

Note that there is no column matrix defined so the equations will be homogeneous. That is, the assembled system equations are

$$\mathbf{St} = \mathbf{0}$$

where S is the direct assembly of S^e and S^b.

Another aspect of interest here is how to post-process the results so as to determine the convective heat loss. Recall from heat transfer that at any point the convective heat loss, dq, is $dq = hP(t - t_\infty)$ where $(t - t_\infty)$ is the surface temperature difference. If we again assume that the heat loss on a typical boundary segment is $t_\infty = 0$

$$Q^b = \int_{\ell^b} dq = h^b P^b \int_{\ell^b} t^b(x)dx \tag{4.76}$$

$$Q^b = h^b P^b \int_{\ell^b} \mathbf{H}^b(x)dx\ \mathbf{t}^b$$

or simply

$$Q^b = 1/2\ h^b P^b \ell^b [1 \quad 1]\mathbf{t}^b$$

$$Q^b = \mathbf{D}^e \mathbf{t}^b . \tag{4.77}$$

Thus if the constant array $\mathbf{D}^b$ is computed and stored for each segment then once the temperatures are computed $\mathbf{t}^b$ can be recovered along with $\mathbf{D}^b$ to compute the loss Q^b. Summing on the total boundary gives the total heat loss. That value would, of course, equal the heat entering at the end x = 0.

As a numerical example let A = 0.01389 ft^2, h = 2 BTU/hr-ft^2 F, K = 120 BTU/hr-ft F, L = 4 ft, P = 0.5 ft, and t_0 = 10 F. The mesh selected for this analysis are shown in Fig. 4.8 along with the results of the finite element analysis given by Akin [1].

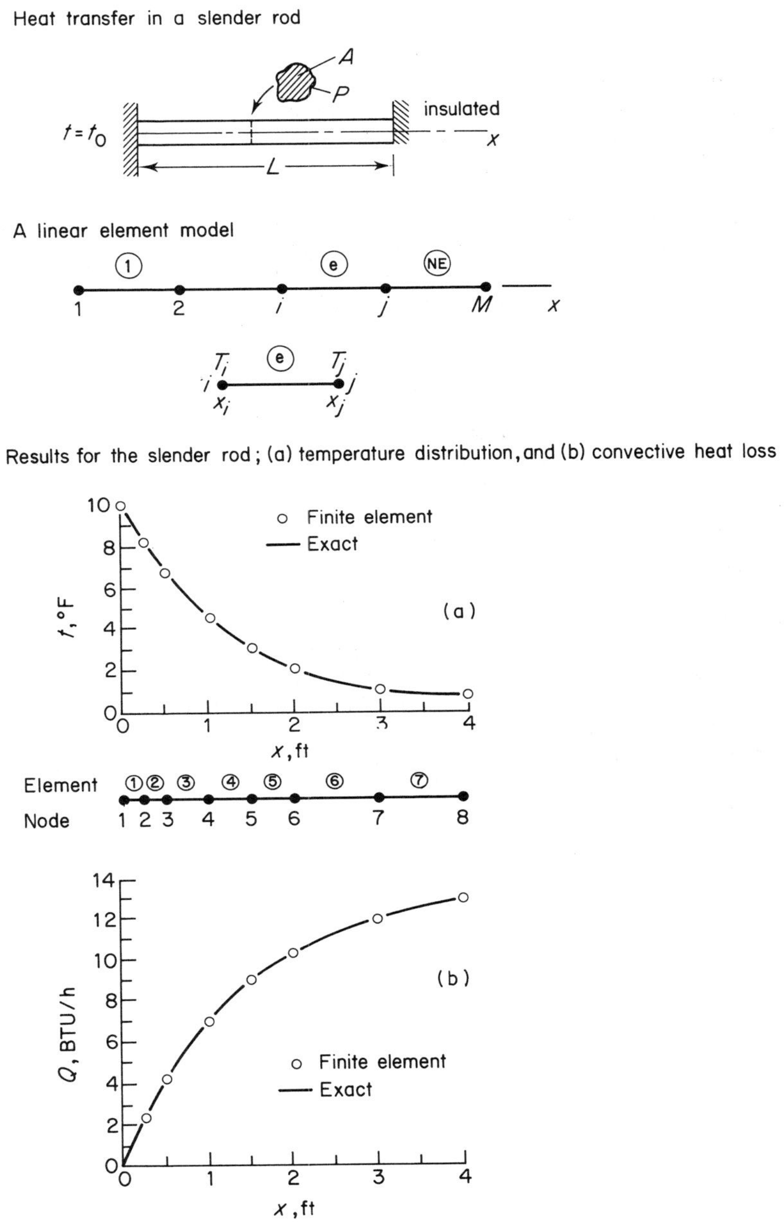

Fig. 4.9 A One-dimensional Heat Transfer Solution

4.7 Exercises

1. As an alternative to the collocation solution in Eq. (4.14) resolve Eq. (4.2) by using points $x_1 = 1/3$ and $x_2 = 2/3$. Verify that the solution is $c_1 = -0.9310$ and $c_2 = 0.3103$.

2. Repeat the collocation solution of Eq. (4.2) using a cubic approximation $y = 1 + c_1x + c_2x^2 + c_3x^3$. Use three collocation points, x_1, x_2, and x_3. Verify that if
   (a) $x_1 = 1/4$, $x_2 = 1/2$ and $x_3 = 3/4$ then $c_1 = -0.9904$, $c_2 = 0.4601$, and $c_3 = -0.1022$.

   (b) While using three Gauss points so $x_1 = 0.1127$, $x_2 = 0.5$, and $x_3 = 0.8873$ gives $c_1 = -0.9948$, $c_2 = 0.4663$, and $c_3 = -0.1036$.

3. Use Eq. (4.19) to solve Eq. (4.2) using four equal length elements.

4. Consider a uniform bar subject to three axial loads. The bar is supported at $x = 0$ and loaded at points $x_a = 3'$, $x_b = 5'$, and $x_c = 6'$. Let the applied loads be $P_a = 2000$ lb, $P_b = 3000$ lb, and $P_c = 5000$ lb, respectively. Assume a uniform stiffness of $AE = 30 \times 10^6$ lb/in. Utilize a three element model to show that the resulting displacements are $u_a = 0.0121$, $u_b = 0.0184$, and $u_c = 0.0204$ in. Also determine the reaction, R.

5. Element Interpolation and Local Coordinates

5.1 Introduction

Up to this point we have relied on the use of a linear interpolation relation that was expressed in **global coordinates** and given by inspection. In the previous chapter we saw numerous uses of these interpolation functions. By introducing more advanced interpolation functions, **H**, we can obtain more accurate solutions. Here we will show how the common interpolation functions are derived. Then a number of expansions will be given without proof. Also, we will introduce the concept of non-dimensional **local** or element, **coordinate** systems. These will help simplify the algebra and make it practical to automate some of the integration procedures.

5.2 Linear Interpolation

Assume that we desire to define a quantity, u, by interpolating in space, from certain given values, **u**. The simplest interpolation would be linear and the simplest space is the line, eg. x-axis. Thus to define u(x) in terms of its values, $\mathbf{u}^e$, at selected points on an element we could choose a linear polynomial in x. That is:

$$u^e(x) = c_1^e + c_2^e x = \mathbf{P}(x)\mathbf{c}^e \tag{5.1}$$

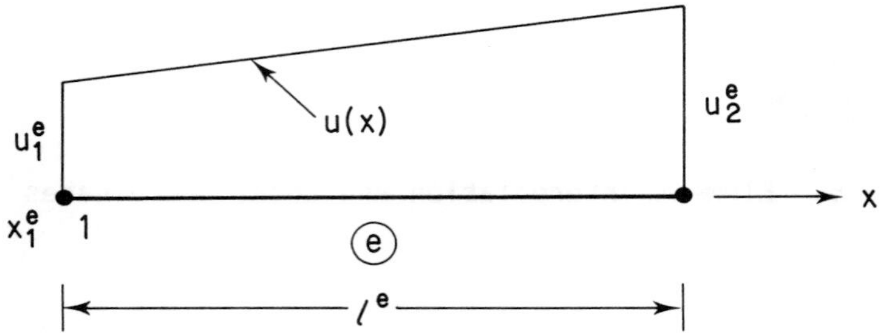

(a) Element in global space

$$u^e(x) = \mathbf{P}(x)\,\mathbf{G}^{e-1}\,\mathbf{u}^e$$

$$\mathbf{u}^{e^T} = \left[u_1{}^e \quad u_2{}^e \right] \sim \text{nodal displacements}$$

$\mathbf{P} \sim$ Behavior in space

$\mathbf{G}^e \sim$ Element geometry

(b) Interpolation in global space

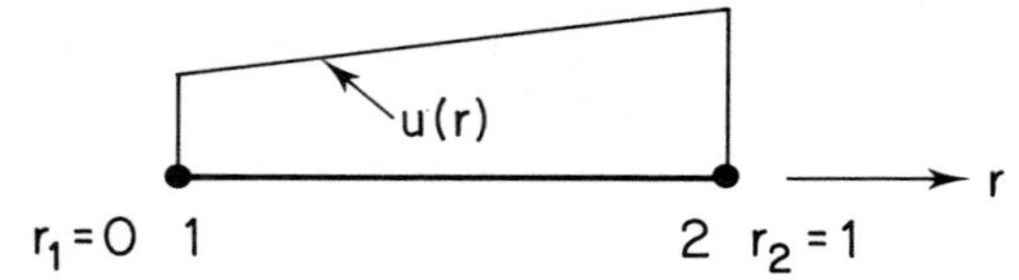

(c) A unit local space

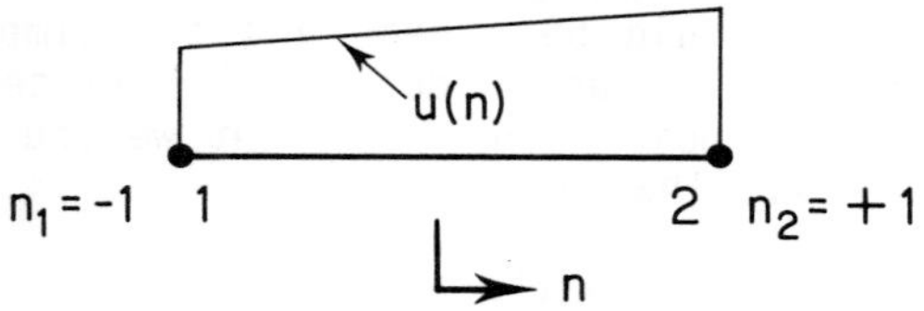

(d) A symmetric local space

Fig. 5.1 One-Dimensional Linear Interpolation

where $\mathbf{P} = [1 \quad x]$ denotes the linear polynomial behavior in space and $\mathbf{c}^{e^T} = [c_1^e \quad c_2^e]$ are undetermined constants that relate to the given values, $\mathbf{u}^e$. Referring to Fig. 5.1 we note that the element has a physical length of ℓ^e and we have defined the nodal values such that $u^e(x_1) = u_1^e$ and $u^e(x_2) = u_2^e$.

To be useful, Eq. (5.1) will be required to be valid at all points on the element, including the nodes. Evaluating Eq. (5.1) at each node of the element gives the set of identities:

$$u^e(x_1^e) = u_1^e = c_1^e + c_2^e x_1^e$$

$$u^e(x_2^e) = u_2^e = c_1^e + c_2^e x_2^e$$

or in matrix form

$$\mathbf{u}^e = \mathbf{G}^e \mathbf{c}^e \tag{5.2}$$

where

$$\mathbf{G}^e = \begin{bmatrix} 1 & x_1^e \\ 1 & x_2^e \end{bmatrix}. \tag{5.3}$$

This shows that the physical constants, $\mathbf{u}^e$, are related to the polynomial constants, $\mathbf{c}^e$, by information on the geometry of the element, $\mathbf{G}^e$. Since $\mathbf{G}^e$ is a square matrix we can (usually) solve Eq. (5.2) to get the polynomial constants:

$$\mathbf{c}^e = \mathbf{G}^{e^{-1}} \mathbf{u}^e \ . \tag{5.4}$$

In this case the element geometry matrix can be easily inverted to give

$$\mathbf{G}^{e^{-1}} = \frac{1}{x_2^e - x_1^e} \begin{bmatrix} x_2^e & -x_1^e \\ -1 & 1 \end{bmatrix} . \tag{5.5}$$

By putting these results into our original assumption, Eq. (5.1), it is possible to write $u^e(x)$ directly in terms of $\mathbf{u}^e$. That is,

$$u^e(x) = \mathbf{P}(x)\mathbf{G}^{e^{-1}}\mathbf{u}^e \tag{5.6}$$

or

$$u^e(x) = [1 \quad x] \frac{1}{\ell^e} \begin{bmatrix} x_2^e & -x_1^e \\ -1 & 1 \end{bmatrix} \begin{Bmatrix} u_1^e \\ u_2^e \end{Bmatrix}$$

$$= \begin{bmatrix} \dfrac{x_2^e - x}{\ell^e} & \dfrac{x - x_1^e}{\ell^e} \end{bmatrix} \begin{Bmatrix} u^e \end{Bmatrix}$$

or simply

$$u^e(x) = \mathbf{H}^e(x)\mathbf{u}^e \tag{5.7}$$

where $\mathbf{H}^e$ is called "the" element interpolation array. Clearly

$$\mathbf{H}^e(x) = \mathbf{P}(x)\mathbf{G}^{e^{-1}} \quad . \tag{5.8}$$

From Eq. (5.6) we can see that the approximate value, $u^e(x)$ depends on the assumed behavior in space, **P**, the element geometry, $\mathbf{G}^e$, and the element nodal parameters, $\mathbf{u}^e$. This is also true in two- and three-dimensional problems.

Since this element interpolation has been defined in a global or physical space the geometry matrix, $\mathbf{G}^e$, and thus $\mathbf{H}^e$ will be different for every element. Of course the algebraic form is common but the numerical terms differ from element to element. For a given type of element it is possible to make **H** unique if a local non-dimensional coordinate is utilized. This will also help reduce the amount of calculus that must be done by hand. Local coordinates are usually selected to range from 0 to 1, or from -1 to +1. These two options are also illustrated in Fig. 5.1. For example consider the **unit coordinates** shown in Fig. 5.1c and let

$$u^e(r) = \mathbf{P}(r)\mathbf{G}^{-1}\mathbf{u}^e \tag{5.9}$$

where the linear polynomial is now $\mathbf{P} = [1 \quad r]$. Repeating the previous steps yields

$$\mathbf{G} = \begin{bmatrix} 1 & 0 \\ 1 & 1 \end{bmatrix} \qquad \mathbf{G}^{-1} = \begin{bmatrix} 1 & 0 \\ -1 & 1 \end{bmatrix}$$

so that

$$u^e(r) = H(r)\mathbf{u}^e \tag{5.10}$$

where the unit coordinate interpolation function is

$$\mathbf{H}(r) = [(1 - r) \quad r] = \mathbf{PG}^{-1}. \tag{5.11}$$

Expanding back to scalar form this means

$$u^e(r) = H_1(r)u_1^e + H_2(r)u_2^e$$

$$= (1 - r)u_1^e + ru_2^e = u_1^e + r(u_2^e - u_1^e)$$

so that at $r = 0$, $u^e(0) = u_1^e$ and at $r = 1$, $u^e(1) = u_2^e$ as required.

A possible problem here is that while this simplifies **H** one may not know "where" a given r point is located in global or physical space. In other words what is x when r is given? One simple way to solve this problem is to note that the nodal values of the global coordinates of the nodes, $\mathbf{x}^e$, are given data. Therefore we can use the concepts in Eq. (5.10) and define

$$x^e(r) = H(r)\mathbf{x}^e \tag{5.12}$$

or

$$x^e(r) = (1 - r)x_1^e + rx_2^e$$

for any r in a given element, e. If we make this popular choice for relating the local and global coordinates we call this an **isoparametric** element. The name implies that a single (iso) set of parametric relations, H(r), is to be used to define the geometry, x(r), as well as the primary unknowns, u(r).

If we select the symmetric, or Gaussian, local coordinates such that $-1 \leq n \leq +1$ then a similar set of interpolation functions are obtained. Specifically

$$u^e(n) = \mathbf{H}(n)\mathbf{u}^e$$

with

$$H_1(n) = (1 - n)/2$$

$$H_2(n) = (1 + n)/2$$

or simply

$$H_i(n) = (1 + n_i n)/2 \tag{5.13}$$

where n_i is the local coordinate of node i. This coordinate system is often called a **natural** coordinate system. Of course, the relation to the global system is

$$x^e(n) = H(n)\mathbf{x}^e. \tag{5.14}$$

The relationship between the unit and natural coordinates is $r = (1 + n)/2$. This will sometimes be useful in converting tabulated data in one system to the other.

The above local coordinates can be used to define how an approximation changes in space. They also allow one to calculate derivatives. For example, from Eq. (5.10)

$$\frac{du^e}{dr} = \frac{d\mathbf{H}(r)}{dr}\,\mathbf{u}^e \tag{5.15}$$

and similarly for other quantities of interest. Another quantity that we will find very important is dx/dr. In a typical linear element, Eq. (5.12) gives

$$\frac{dx^e(r)}{dr} = \frac{dH_1}{dr}\,x_1{}^e + \frac{dH_2}{dr}\,x_2{}^e$$

$$= -x_1{}^e + x_2{}^e$$

or simply $dx^e/dr = \ell^e$. By way of comparison, if the natural coordinate is utilized

$$\frac{dx^e(n)}{dn} = \ell^e/2. \tag{5.16}$$

This illustrates that the choice of the local coordinates has more effect on the derivatives than it does on the interpolation itself.

The use of unit coordinates is more popular with **simplex elements.** These are elements where the number of nodes is one higher than the dimension of the space. The generalization of unit coordinates for common simplex elements is illustrated in Fig. 5.2. For simplex elements the natural coordinates become **area coordinates** and **volume coordinates** which the author finds rather unnatural. Both unit and natural coordinates are effective for use on squares or cubes in the local space. In global space these shapes become quadrilaterals or hexahedra as illustrated in Fig. 5.3. Generally the natural coordinates are more popular for these shapes.

5.3 Quadratic Interpolation

The next logical spatial form to pick is that of a quadratic polynomial. Select three nodes on the line element. Two are at the ends and the third is inside the element. In local space the third node is at the element center. Thus the local unit coordinates are $r_1 = 0$, $r_3 = 1/2$, and $r_2 = 1$. Usually it is desirable to have x_3 also at the center of the element in global space.

If we repeat the previous procedure using $u(r) = c_1 + c_2 r + c_3 r^2$ then the element interpolation functions are found to be

$$\begin{aligned} H_1(r) &= 1 - 3r + 2r^2 \\ H_2(r) &= -r + 2r^2 \\ H_3(r) &= 4r - 4r^2 \\ \hline \Sigma H_i(r) &= 1 \end{aligned} \tag{5.17}$$

These quadratic functions are completely different from the linear functions. Note that these functions have a sum that is unity at any point, r, in the element. These three functions are plotted in Fig. 5.4. That figure illustrates another common feature of interpolation functions. They are unity at one node and zero at all others. Specifically

$$H_i(r_j) = \delta_{ij} = \begin{cases} 1 & \text{if } i = j \\ 0 & \text{if } i \neq j. \end{cases}$$

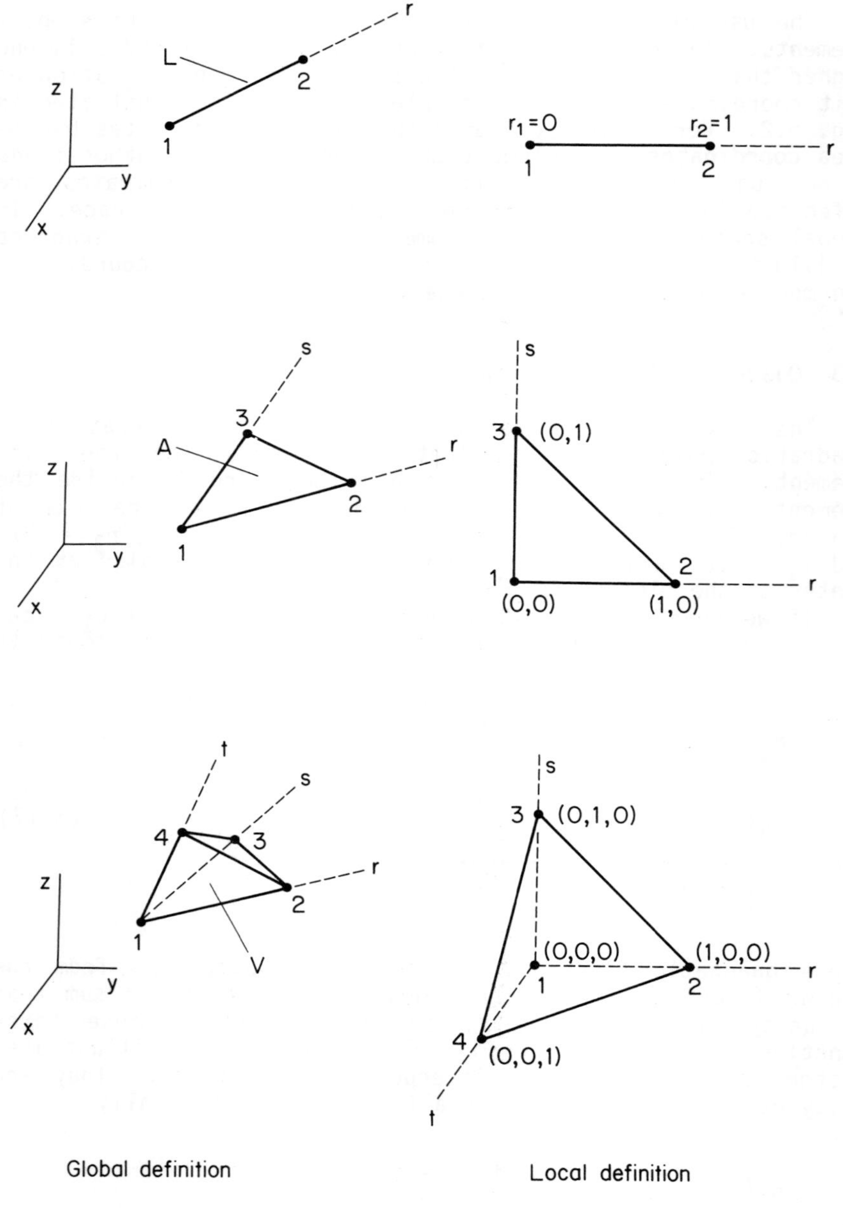

Fig. 5.2 The Simplex Element Family

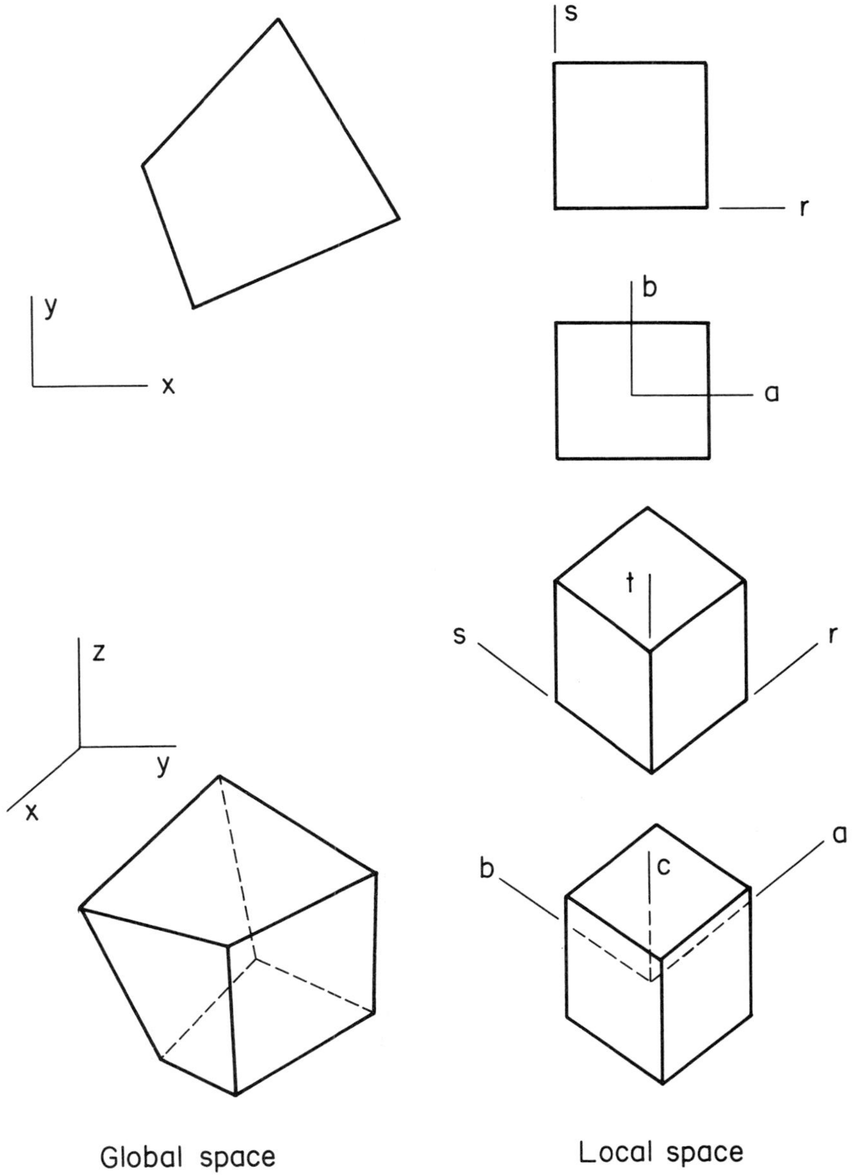

Fig. 5.3 Global and Local Spaces for Quadrilaterals and Hexahedra

If these functions are expressed in natural coordinates then $n_1 = -1$, $n_2 = 1$, and $n_3 = 0$ and the functions are

$$H_1(n) = n(n - 1)/2$$

$$H_2(n) = n(n + 1)/2$$

$$H_3(n) = 1 - n^2. \tag{5.18}$$

5.4 Lagrange Interpolation

Clearly this one dimensional procedure can be readily extended by adding more nodes to the interior of the element. Usually the additional nodes are equally spaced along the element. However, they can be placed in arbitrary locations. The interpolation function for such an element is known as the Lagrange interpolation polynomial. The one-dimensional m-th order **Lagrange interpolation** polynomial is the ratio of two products. For an element with (m + 1) nodes, r_i, i = 1,2,..., (m+1), the interpolation function for the k-th node is

$$L_k{}^m(n) = \prod_{\substack{i=1 \\ i \neq k}}^{m+1} (n - n_i) \Big/ \prod_{\substack{i=1 \\ i \neq k}}^{m+1} (n_k - n_i) \quad .$$

This is a complete m-th order polynomial in one dimension. It has the property that $L_k(n_i) = \delta_{ik}$. For local coordinates given on the domain [-1, 1] a typical term for three equally spaced nodes is

$$L_3(n) = \frac{(n - (-1))(n - 1)}{(0 - (-1))(0 - 1)} = (1 - n^2).$$

Similarly $L_1(n) = n(n - 1)/2$ and $L_2 = n(n + 1)/2$ and their sum is unity. Figure 5.5 shows typical node locations and interpolation functions for members of this family of **complete polynomial** functions on simplex elements.

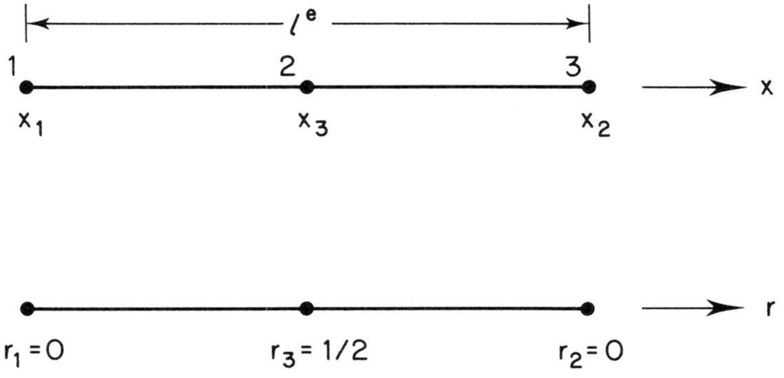

(a) Global and local node locations

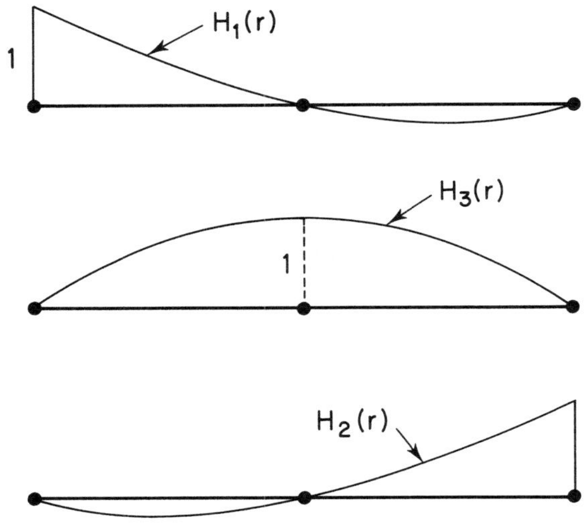

(b) Element interpolation functions

Fig. 5.4 The Quadratic Line Element

5.5 **Hermitian Interpolation**

All of the interpolation functions considered so far have c^0 continuity between elements. That is, the function being approximated is continuous between elements but its derivative is discontinuous. However, we have seen from Eq. (4.41) that some applications, such as a beam analysis, also require that the derivative be continuous. These c^1 functions are most easily generated by using derivatives, or slopes, as nodal degrees of freedom.

The simplest element in this family is the two node line element where both y and dy/dx are taken as nodal degrees of freedom. Note that a global derivative has been selected as a degree of freedom. Since there are two nodes with two dof each, the interpolation function has four constants. Thus, it is a cubic polynomial. The form of this **Hermite polynomial** is well known. The element is shown in Fig. 5.6 along with the interpolation functions and their global derivatives. The latter quantities are obtained from the relation between local and global coordinates, eg., Eq. (5.16).

On rare occasions one may also need to have the second derivatives continuous between elements. Typical c^2 equations of this type are also given in Fig. 5.6 and elsewhere. Since derivatives have also been introduced as nodal parameters the previous statement that $\Sigma H_i = 1$ is no longer true.

5.6 **Hierarchical Interpolation**

Recently some alternate types of interpolation have become popular. They are called **hierarchical functions.** The unique feature of these polynomials is that the higher order polynomials contain the lower order ones. Thus to get new functions you simply add some terms to the old functions. To illustrate this concept let us return to the linear element in local unit coordinates. In that element

$$u^e(r) = H_1(r)u_1^e + H_2(r)u_2^e$$

$-1 < n < 1$	$0 < r < 1$
A) LINEAR	
$H_1 = (1 - n)/2$	$H_1 = (1 - r)$
$H_2 = (1 + n)/2$	$H_2 = r$
B) QUADRATIC	
$H_1 = n(n - 1)/2$	$H_1 = (r - 1)(2r - 1)$
$H_2 = n(n + 1)/2$	$H_2 = r(2r - 1)$
$H_3 = (1 + n)(1 - n)$	$H_3 = 4r(1 - r)$
C) CUBIC	
$H_1 = (1 - n)(3n + 1)(3n - 1)/16$	$H_1 = (1 - r)(r - 3r)(1 - 3r)/2$
$H_2 = (1 + n)(3n + 1)(3n - 1)/16$	$H_2 = r(2 - 3r)(1 - 3r)/2$
$H_3 = 9(1 + n)(n - 1)(3n - 1)/16$	$H_3 = 9r(1 - r)(2 - 3r)/2$
$H_4 = 9(1 + n)(1 - n)(3n + 1)/16$	$H_4 = 9r(1 - r)(3r - 1)/2$

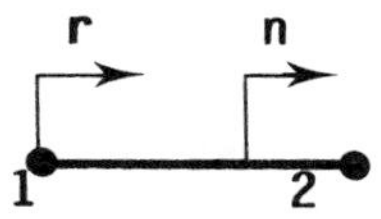

Fig. 5.5 Typical Lagrange Interpolations

$x = Lr$

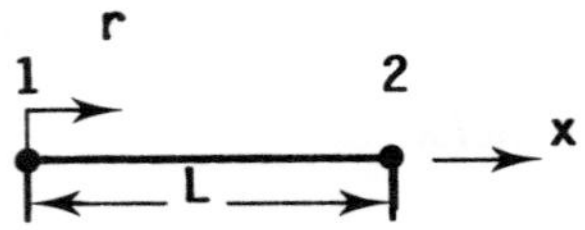

$()' = d()/dx$

A) C^1: $U = H_1U_1 + H_2U_1' + H_3U_2 + H_4U_2'$

$$H_1(r) = (2r^3 \quad 3r^2 + 1)$$
$$H_2(r) = (r^3 - 2r^2 + r)L$$
$$H_3(r) = (3r^2 - 2r^3)$$
$$H_4(r) = (r^3 - r^2)L$$

B) C^2: $U = H_1U_1 = H_2U_1' + H_3U'' + H_4U_2 + H_5U_2' + H_6U_2''$

$$H_1 = (1 - 10r^3 + 15r^4 - 6r^5)$$
$$H_2 = (r - 6r^3 + 8r^4 - 3r^5)L$$
$$H_3 = (r^2 - 3r^3 + 3r^4 - r^5)L^2/2$$
$$H_4 = (10r^3 - 15r^4 + 6r^5)$$
$$H_5 = (7r^4 - 3r^5 - 4r^3)L$$
$$H_6 = (r^3 - 2r^4 + r^5)L^2/2$$

C) C^3: $U = H_1U_1 + H_2U_1' + H_3U_1'' + H_4U_1'''$
$+ H_5U_2 + H_6U_2' + H_7U_2'' + H_8U_2'''$

$$H_1 = (1 - 35r^4 + 84r^5 - 70r^6 + 20r^7)$$
$$H_2 = (r - 20r^4 + 45r^5 - 36r^6 + 10r^7)/L^2/2$$
$$H_3 = (r^2 - 10r^4 + 20r^5 - 15r^6 + 4r^7)L^2/2$$
$$H_4 = (r^3 - 4r^4 + 6r^5 - 4r^6 + r^7)L^3/6$$
$$H_5 = (35r^4 - 84r^5 + 70r^6 - 20r^7)$$
$$H_6 = (10r^7 - 34r^6 + 39r^5 - 15r^4)L$$
$$H_7 = (5r^4 - 14r^5 + 13r^6 - 4r^7)L^2/2$$
$$H_8 = (r^7 - 3r^6 + 3r^5 - r^4)L^3/6$$

Fig. 5.6 Hermitian Interpolation in Unit Coordinates

where the two H_i are given in Eq. (5.11). We want to generate a quadratic interpolation form that will not destroy these H_i as Eq. (5.17) did. The key to accomplishing this goal is to note that the second derivative of Eq. (5.11) is everywhere zero. Thus, if we introduce an additional degree of freedom related to the second derivative of u it will not affect the linear terms. Figure 5.7 shows the linear element where we have added a third control node to be associated with the quadratic additions. At the third node let the degree of freedom be the second local derivative, d^2u/dr^2. Upgrade the approximation by setting

$$u(r) = H_1(r)u_1^e + H_2(r)u_2^e + Q_3(r)\frac{d^2u^e}{dr^2} \tag{5.19}$$

where the hierarchical quadratic addition is:

$$Q_3(r) = c_1 + c_2r + c_3r^2.$$

The three constants are found from the conditions that it vanishes at the two original nodes, so as not to change H_1 and H_2, and the second derivative is unity at the new node. The result is

$$Q_3(r) = (r^2 - r)/2 \quad . \tag{5.20}$$

The higher order hierarchical functions are becoming increasingly popular. They utilize the higher derivatives at the center node. These functions must vanish at the end nodes. Most of their derivatives, excluding the first, must vanish at the center node. Finally, the function $H_p(n)$, $p \geqslant 2$ must have its p-th derivative take on a value of unity at the center node. The resulting functions are not unique. A common set of hierarchical functions are

$$H_p(n) = (n^p - b)/p!, \; p \geqslant 2 \tag{5.21}$$

where

$$b = \begin{cases} 1 & \text{if } p \text{ is even} \\ n & \text{if } p \text{ is odd.} \end{cases}$$

This relation is zero at the ends, $n = \pm 1$. The first derivative of these functions are

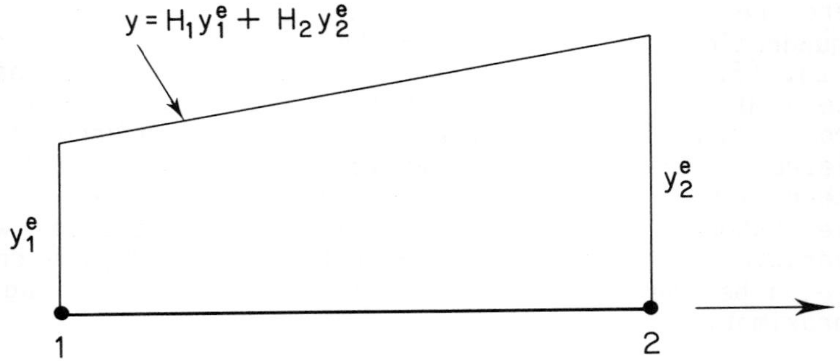

(a) Standard linear interpolation

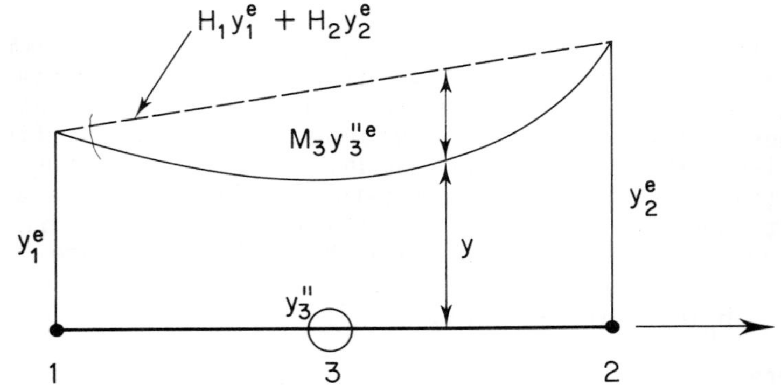

(b) Standard linear function with hierarchical quadratic addition

Fig. 5.7 A Quadratic Hierarchical Element

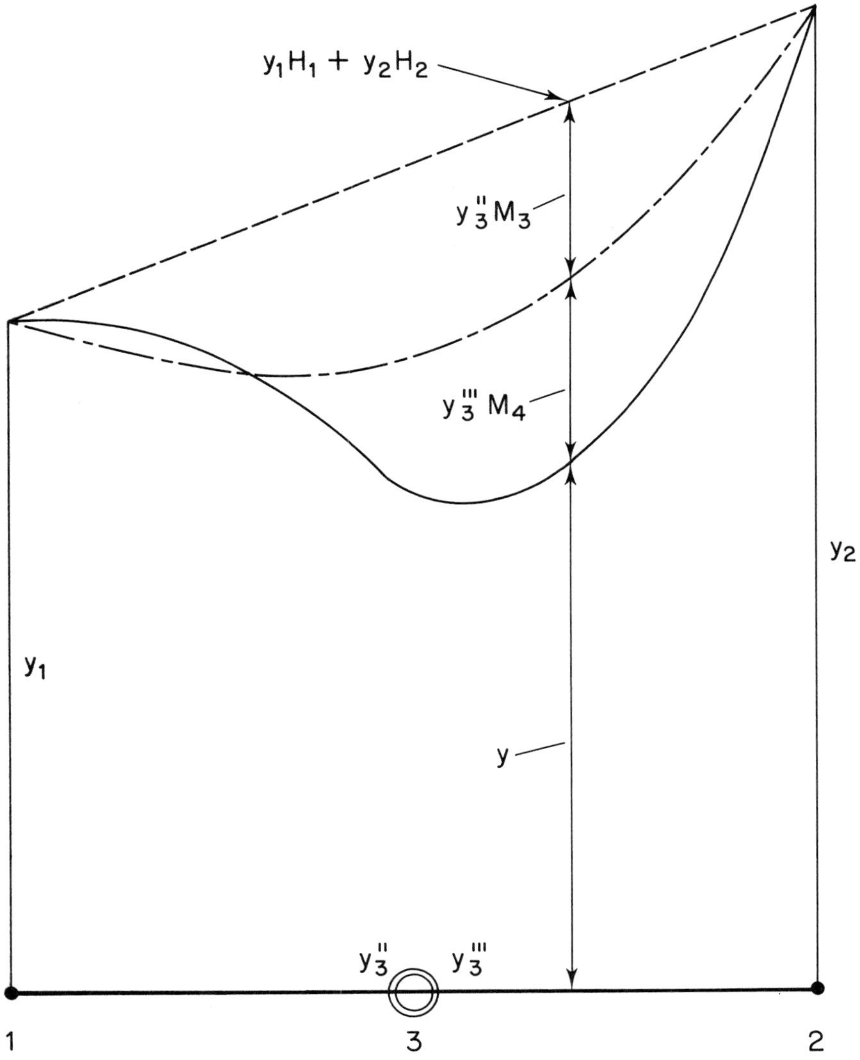

Fig. 5.8 A Typical Cubic al Function

$$H'_p = (pn^{(p-1)} - b')/p!$$

and since b" = 0 the second derivative is

$$H'' = p(p-1)n^{(p-2)}/p! = n^{(p-2)}/(p-2)!.$$

Proceeding in this manner it is easy to show by induction that the m-th derivative, for $m \geqslant 2$, is

$$H_p^{(m)}(n) = n^{(p-m)}/(p-m)! \quad . \tag{5.22}$$

At the center point, n = 0, the derivative has a value of

$$H_p^{(m)}(0) = \begin{cases} 0 & \text{if } m \neq p \\ 1 & \text{if } m = p. \end{cases}$$

We will see later that when hierarchical functions are utilized the element matrices for a p-th order polynomial are partitions of the element matrices for a (p + 1) order polynomial. A typical cubic element is given in Fig. 5.8.

5.7 Exercises

1. Verify Eq. (5.16)
2. Verify that the second derivatives of the cubic Hermite functions are

$$H''_1 = (12r - 6)/L^2 \qquad H''_3 = (6 - 12r)/L^2$$
$$H''_2 = (6r - 4)/L \qquad H''_4 = (6r - 2)/L.$$

3. Utilize the Lagrange interpolation rule to verify the cubic functions in Fig. 5.5 in: a) Unit coordinates, b) Natural Coordinates.
4. Write subroutines to evaluate the interpolation functions for the unit coordinate quadratic line element and their derivatives (say SHP3L and DER3L).
5. Write subroutines to evaluate the interpolation functions and first two derivatives for the cubic Hermite element in unit coordinates. (say SHPCU, DERCU, and DER2CU).

6. Exact and Numerical Integration in 1-D

6.1 Introduction

Since the finite element method is based on integral relations it is logical to expect that one should strive to carry out the integrations as efficiently as possible. In some cases we will employ exact integration. In other cases we may find that the integrals can become too complicated to integrate exactly. In such cases the use of numerical integration will prove useful or essential. The important topics of local coordinate integration and Gaussian quadratures will be introduced here. They will prove useful when dealing with higher order interpolation functions in complicated element integrals.

6.2 Local Coordinate Jacobian

We have previously seen that the utilization of local element coordinates can greatly reduce the algebra required to establish a set of interpolation functions. Later we will see that some 2-D elements must be formulated in local coordinates in order to meet the interelement continuity requirements. Thus we should expect to often encounter local coordinate interpolation. However, the governing integral expressions must be evaluated with respect to a unique global or physical coordinate system. Clearly these two coordinate systems must be related. The relationship for integration with a change of variable (change of coordinate) was defined in elementary concepts from calculus. At this point it would be useful to review these concepts from calculus. Consider a definite integral

$$I = \int_a^b f(x)dx, \quad a < x < b \tag{6.1}$$

where a new variable of integration, r, is to be introduced such that $x = x(r)$. Here it is required that the function $x(r)$ is continuous and have a continuous derivative in the interval $\alpha \leq r \leq \beta$. The region of r directly corresponds to the region of x such that when r varies between α and β then x varies between $a = x(\alpha)$ and $b = x(\beta)$. In that case

$$I = \int_a^b f(x)dx = \int_\alpha^\beta f(x(r))\frac{dx}{dr}\, dr \tag{6.2}$$

or

$$I = \int_\alpha^\beta f(r)\ J\ dr \tag{6.3}$$

where

$$J = dx/dr \tag{6.4}$$

is called the **Jacobian** of the coordinate transformation.

6.3 Exact Polynomial Integration

If we utilize the unit coordinates then $\alpha = 0$ and $\beta = 1$. Then from Section (5.2) the Jacobian is $J = \ell^e$ in an element domain defined by linear interpolation. By way of comparison if one employs natural coordinates then $\alpha = -1$, $\beta = +1$, and from Eq. (5.16) $J = \ell^e/2$.

Generally we will use interpolation functions that are polynomials. Thus the element integrals of them and/or their derivatives will also contain polynomial terms. Therefore, it will be useful to consider expressions related to typical polynomial terms. A typical polynomial term is r^m where m is an integer. Thus from the above

$$I = \int_{x_1^e}^{x_2^e} r^m dx = \int_0^1 r^m\ \ell^e dr$$

$$I = \ell^e \int_0^1 r^m dr = \ell^e \left. \frac{r^{(1+m)}}{1+m} \right|_0^1 = \ell^e/(1+m) \;. \tag{6.5}$$

A similar expression can be developed for the natural coordinates. Later we will tabulate the extension of these concepts to two- and three-dimensional integrals.

As an example of the use of Eq. (6-5) consider the integration in Eq. (4.75):

$$I = \int_{\ell^e} \mathbf{H}^{e^T} \mathbf{H}^e dx \;. \tag{6.6}$$

Recall that the integral of a matrix is the matrix resulting from the integration of each of the elements of the original matrix. If linear interpolation is selected for $\mathbf{H}^e$ then typical terms will include $H_1{}^2$, H_1H_2, etc. Thus one obtains:

$$I_{11} = \int_{\ell^e} H_1{}^2(r)dx = \int_{\ell^e} (1-r)^2 dx$$

$$= \int_{\ell^e} (1 - 2r + r^2)dx = \ell^e(1 - 2/2 + 1/3) = \ell^e/3$$

$$I_{12} = \int_{\ell^e} H_1H_2 dx = \int_{\ell^e} (1-r)r dx = \ell^e(1/2 - 1/3)$$

$$= \ell^e/6$$

and

$$I_{22} = \int_{\ell^e} H_2{}^2 dx = \int_{\ell^e} r^2 dx = \ell^e/3$$

so that

$$I = \frac{\ell^e}{6}\begin{bmatrix} 2 & 1 \\ 1 & 2 \end{bmatrix} \;. \tag{6.7}$$

Similarly if one employs the quadratic $\mathbf{H}$ in Eq. (5-17) one obtains:

$$I = \frac{\ell^e}{30}\begin{bmatrix} 4 & -1 & 2 \\ -1 & 4 & 2 \\ 2 & 2 & 16 \end{bmatrix} \;. \tag{6.8}$$

By way of comparison if one selects the hierarchical quadratic polynomial in Eq. (5.19) the above integral becomes

$$I = \frac{\ell^e}{6}\left[\begin{array}{cc:c} 2 & 1 & -1/4 \\ 1 & 2 & -1/4 \\ \hdashline -1/4 & -1/4 & 1/10 \end{array}\right] .$$

Note that the top left portion of this equation is the same as Eq. (6.7) which was obtained from the linear polynomial. This desirable feature of hierarchical elements was mentioned in Sec. 5.6.

6.4 Numerical Integration

To continue on our goal of automating the finite element analysis procedure it is desirable to carry out the required integrations numerically. One could easily spend many pages discussing one-dimensional numerical integration. There are numerous common integration rules such as the trapezoidal rule, Simpson's rule, Newton-Cotes formula, etc. However, our major interest is to integrate polynomials in the most accurate and efficient manner. Both of the latter points suggest the use of **Gaussian quadrature** rules, or Lobatto rules.

Since the finite element method requires a large amount of integration, it is imperative that one obtain the greatest possible accuracy with the minimum cost (computer time). The most accurate numerical method in ordinary use for integrating polynomials is the **Gauss quadrature formula.** Consider the definite integral

$$I = \int_a^b f(x)\, dx, \tag{6.9}$$

which is to be computed numerically from a given number, n, of values of f(x). Gauss considered the problem of determining which values of x should be chosen in order to get the greatest possible accuracy. In other words, how shall the interval (a,b) be subdivided so as to give the best possible results? Gauss found that the "n" points in the interval should not be equally spaced but should be symmetrically placed with respect to the mid-point of the interval. Some results of Gauss's work are outlined below. Let y denote f(x) in the integral to be computed. Define a change of variable

$$x(r) = 1/2\ (b - a)r + 1/2\ (b + a) \tag{6.10}$$

so that the non-dimensional limits of integration of r become -1 and +1. The new value of y(r) is

$$y = f(x) = f\ [1/2\ (b - a)r + 1/2\ (b + a)] = \Phi(r). \tag{6.11}$$

Noting from Eq. (6.10) that dx = 1/2 (b - a)dr, the original integral becomes

$$I = 1/2\ (b - a) \int_{-1}^{1} \Phi(r)dr. \tag{6.12}$$

Gauss showed that the integral in Eq. (6.12) is given by

$$\int_{-1}^{1} \Phi(r)dr = \sum_{i=1}^{n} W_i \Phi(r_i),$$

where W_i and r_i represent tabulated values of the **weight functions** and **abscissae** associated with the n points in the non-dimensional interval (-1, 1). Thus the final result is

$$I = 1/2\ (b - a) \sum_{i=1}^{n} W_i \Phi(r_i) = \sum_{i=1}^{n} f(x(r_i))W_i. \tag{6.13}$$

Gauss also showed that this equation will exactly integrate a polynomial of degree (2n - 1). For a higher number of space dimensions (which range from -1 to +1) one obtains a multiple summation.

Since Gaussian quadrature data are often tabulated in references for the range $-1 \leq r \leq +1$ it is popular to use the natural coordinates in defining element integrals. However, one can convert the tabulated data to any convenient system such as the unit coordinate system where $0 \leq r \leq 1$. The latter may be more useful on triangular regions.

As an example of Gaussian quadratures, consider the following one-dimensional integral:

$$\mathbf{I} = \int_{1}^{2} \begin{bmatrix} 2 & 2x \\ 2x & (1+2x^2) \end{bmatrix} dx = \int_{1}^{2} \mathbf{F}(x)\ dx.$$

If two Gauss points are selected, then the tabulated values from Table VI.I give $W_1 = W_2 = 1$; and $r_1 = 0.57735 = -r_2$. The change of variable gives $x(r) = (r + 3)/2$, so that $x(r_1) = 1.788675$ and $x(r_2) = 1.211325$. Therefore, from Eqn. (6.14)

$$I = 1/2\ (2 - 1)\ [\ W_1F(x(r_1)) + W_2F(x(r_2))\ ]$$

$$= 1/2\ (1) \left((1) \begin{bmatrix} 2 & 2(1.788675) \\ \text{sym.} & 1+2(1.788675)^2 \end{bmatrix} \right.$$

$$\left. + (1) \begin{bmatrix} 2 & 2(1.211325) \\ \text{sym.} & 1+2(1.211325)^2 \end{bmatrix} \right)$$

$$I = \begin{bmatrix} 2.00000 & 3.00000 \\ 3.00000 & 5.66667 \end{bmatrix} ,$$

which is easily shown to be in good agreement with the exact solution.

As another example consider a typical term in Eq. (6.8). Specifically, from Eqs. (5.16) and (5.18)

$$I_{33} = \int_{\ell^e} H_3^2\ dx = \ell^e/2 \int_{-1}^{+1} H_3^2(n)dn$$

$$= \ell^e/2 \int_{-1}^{+1} (1 - n^2)^2 dn \quad .$$

Since the polynomial terms to be integrated are fourth order we should select $(2m - 1) = 4$, or $m = 3$ Gaussian points. Then

$$I_{33} = \ell^e/2 \sum_{i=1}^{3} (1 - n_i^2)^2\ W_i$$

$$I_{33} = \ell^e/2\ (0.55556(1.00000 - (0.77459)^2)^2$$
$$+ 0.88889(1 - (0)^2)^2$$
$$+ 0.55556(1.00000 - (+0.77459)^2)^2)$$

$$I_{33} = \ell^e/2\ (0.08889 + 0.88889 + 0.08889) = 0.5333\ell^e$$

Table VI.I Gaussian Quadrature Data: $\int_a^b f(r)\,dr = \sum_{i=1}^{n} f(r_i)W_i$

	Natural Coordinates, a = -1, b = +1		Unit Coordinates, a = 0, b = 1	
n	$\pm r_i$	W_i	r_i	W_i
1	0	2	1/2	1
2	$\pm 1/\sqrt{3}$	1	$(\sqrt{3}\pm 1)/2\sqrt{3}$	1/2
3	$\pm\sqrt{15}/5$	5/9	$(5\pm\sqrt{15})/10$	5/18
	0	8/9	1/2	4/9
4	$\pm\sqrt{\frac{3+\sqrt{24/5}}{7}}$	$\frac{18-\sqrt{30}}{36}$		
	$\pm\sqrt{\frac{3-\sqrt{24/5}}{7}}$	$\frac{18+\sqrt{30}}{36}$		

Table VI.II Lobatto Quadrature Data: $\int_a^b f(r)\,dr = \sum_{i=1}^{n} f(r_i)W_i$

Natural Coordinates, a = -1, b = +1 Unit Coordinates, a = 0, b = 1

n	r_i	W_i	r_i	W_i
2	± 1	1	0	1/2
			1	1/2
3	± 1	1/3	0	1/6
	0	4/3	1/2	4/6
			1	1/6
4	± 1	1/6	0	1/12
	$\pm 1/\sqrt{5}$	5/6	$(1 \pm 1/\sqrt{5})/2$	5/12
			1	1/12
5	± 1	1/10	0	1/20
	$\pm 3/\sqrt{7}$	49/90	$(1 \pm \sqrt{3}/\sqrt{7})/2$	49/180
	0	32/45	1/2	32/90
			1	1/20

which agrees well with the exact value of $16\ell^e/30$ given in Eq. (6.8). Some typical quadrature data for lines are shown in Table VI.I and VI.II.

6.5 Exercises

1. Write a subroutine, GAUSCO, that will return the tabulated values of the Gaussian weights and abscissae for a given number of points, n.
2. Write a subroutine, LOBATO, that will return the weights and abscissae for a given number of points, n.
3. Repeat the solution of the stepped bar system in Fig. 4.7. Use two quadratic line elements in unit coordinates. Let the new midpoint nodes be numbered as points 4 and 5. Then A) Calculate the local derivatives of **H**. Verify that the element Jacobian is a constant. Calculate the global derivatives of **H**. B) Utilize Eq. (6.5) to verify that an element stiffness is

$$S^e = \frac{EA}{3L}\begin{bmatrix} 7 & 1 & -8 \\ 1 & 7 & -8 \\ -8 & -8 & 16 \end{bmatrix}.$$

   C) For a constant body force X verify that the element vector is $\mathbf{C}^{e^T} = AXL\,[1 \quad 1 \quad 4]/6$.
   D) Employ the two element solution to verify that the nodal displacements are $[0.0 \quad 1.565 \quad 3.939 \quad 0.803 \quad 2.769] \times 10^{-2}$ in.
   E) Evaluate and sketch the linear stress distribution in each element and compare with Fig. 4.6.
4. Integrate $f(r) = (1 - r - 2r^2)$ from -1 to 1 by using a) calculus, b) Gaussian quadrature, c) Lobatto quadrature.
5. Consider the linear strain bar element in problem 3. Apply a rigid body displacement given by $u_1 = u_2 = u_3 = C$, where C is an arbitrary constant. Calculate the strain in an element due to these displacements.
6. Integrate $f(x) = (1 + x + x^2)$ from 1 to 2 by using a) calculus, b) Gaussian quadrature.

TRUSS ELEMENTS AND COORDINATE TRANSFORMATION

7.1 Introduction

The truss element is a very common structural member. Recall that a truss element is a two force member. That is, it is loaded by two equal and opposite collinear forces. These two forces act along the line through the two connection points of the member. In elementary statics we compute the forces in truss elements as if they were rigid bodies. However, there was a class of problems, called statically indeterminant, that could not be solved by treating the members as rigid bodies. With the finite element approach we will be able to solve both classes of problems.

In Section 4.5 the equilibrium equation for an elastic bar was developed. Clearly, the elastic bar is a special form of a truss member. To extend the previous work to include trusses in two- or three-dimensions basically requires some review of analytic geometry. Thus we begin by reviewing that subject.

7.2 Direction Cosines

Consider a directed line segment in global space going from point 1 at (x_1, y_1, z_1) to point 2 at (x_2, y_2, z_2). Then the length of the line between the two points has components parallel to the axes of

$$\ell_x = x_2 - x_1$$

$$\ell_y = y_2 - y_1$$

$$\ell_z = z_2 - z_1 \tag{7.1}$$

and the total length is $\ell^2 = (\ell_x^{\ 2} + \ell_y^{\ 2} + \ell_z^{\ 2})$. Specifying the end points of a line is a common way of locating its direction in space. Another common way to describe the direction is to give the **direction angles** or the corresponding **direction cosines.** Let the direction angles from the x-, y-, and z-axes to the line segment be denoted by ϕ_x, ϕ_y, and ϕ_z, respectively. Recall the relation between the total magnitude of a vector and its components, ie., $\ell_x = \ell \operatorname{Cos} \phi_x$, etc. We generally will find the inverse geometric relation more useful. Specifically

$$\operatorname{Cos} \phi_x = \ell_x / \ell$$

$$\operatorname{Cos} \phi_y = \ell_y / \ell$$

$$\operatorname{Cos} \phi_z = \ell_z / \ell \ . \tag{7.2}$$

For two-dimensional problems we will assume that the structure lies in the global x-y plane so that $\ell_z = 0$, $\operatorname{Cos} \phi_z = 0$, and $\phi_z = 90$. In that special case only one angle is required to describe the direction rather than the usual three. It is common then to select ϕ_x as the required angle and to omit reference to $\phi_y = 90 - \phi_x$ and to replace the second direction cosine with the relation

$$\operatorname{Cos} \phi_y = \operatorname{Sin} \phi_x \qquad (\text{for } \phi_z = 90). \tag{7.3}$$

This is illustrated in Fig. 7.1. For two-dimensional problems one can utilize the simplicity of referring to a single angle. However, if one wants to automate the analysis for two- and three-dimensional problems then it is best in the long run to refer to the direction cosines.

7.3 Transformation of Displacement Components

To extend the bar element to a general truss element we need to consider the relations between a local coordinate system that is parallel and perpendicular to the element and the fixed global coordinate directions. Let the local x-axis lie along the member. That is, it passes through the two end points of the member. This means that the direction cosines of the local x-axis are the same as those for the line segment.

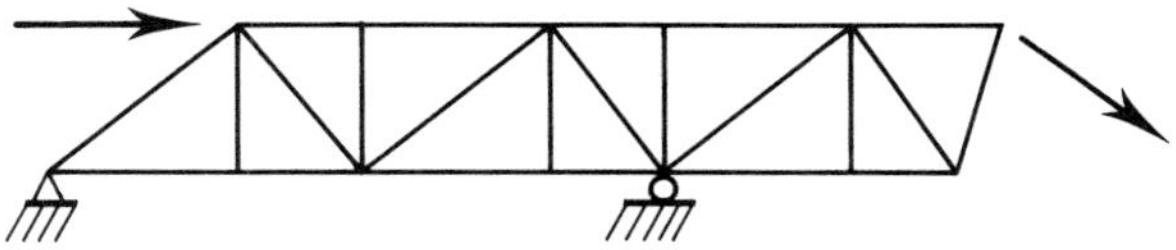

(a) A typical truss system

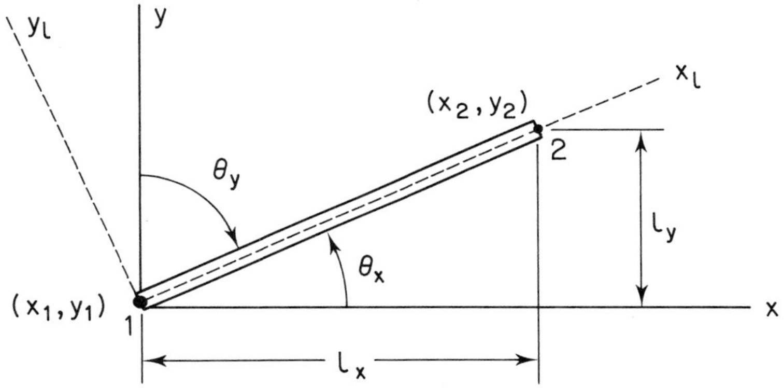

(b) Location of a typical element in local and global coordinates

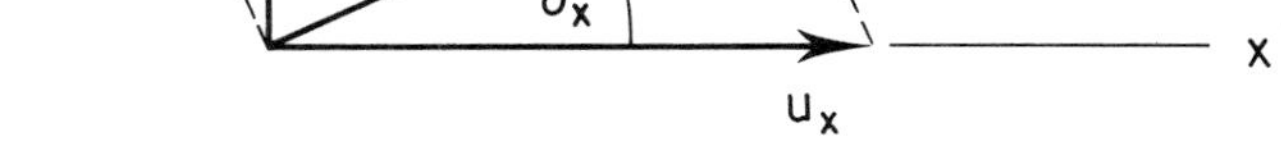

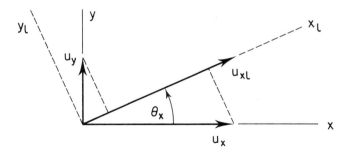

(c) Relation between local and global displacement components

Fig. 7.1 A Truss Structure

The bar element had a single displacement, u, at any point. Now that local displacement vector will have components in the global space. Let the global displacements of a point be denoted by u_x, u_y, and u_z. To be consistent with this, one could also define three local components of the displacement. For a bar element the local y- and z-components are identically zero. Later we will consider members that have no zero local components. Thus we will consider the general case of transformation of local displacement components.

Referring to Fig. 7.1c one finds from geometry that the local x displacement is related to the two-dimensional global displacements by

$$u_{x\ell} = u_{xg} \text{ Cos } \theta_x + u_{yg} \text{ Cos } \theta_y.$$

Similarly, if there was a local y-component of displacement it would be related to the global components by

$$u_{y\ell} = - u_{xg} \text{ Cos } \theta_y + u_{yg} \text{ Cos } \theta_x.$$

Writing these identities in a matrix form in terms of $\theta_x = \theta$

$$\begin{Bmatrix} u_x \\ u_y \end{Bmatrix}_\ell = \begin{bmatrix} \text{Cos}\theta & \text{Sin}\theta \\ -\text{Sin}\theta & \text{Cos}\theta \end{bmatrix} \begin{Bmatrix} u_x \\ u_y \end{Bmatrix}_g \tag{7.4}$$

or symbolically this **transformation** is

$$\mathbf{u}_\ell = \mathbf{t}(\theta)\mathbf{u}_g$$

where **t** is a nodal **transformation matrix** and $\mathbf{u}_g$ and $\mathbf{u}_\ell$ denote the global and local displacement components, respectively, at a point. If this relation is written at each node of the element it defines the element dof transformation matrix, T. Specifically

$$\begin{Bmatrix} u_{1x} \\ u_{1y} \\ \hline u_{2x} \\ u_{2y} \end{Bmatrix}_\ell^e = \left[\begin{array}{cccc} \text{Cos}\theta & \text{Sin}\theta & 0 & 0 \\ -\text{Sin}\theta & \text{Cos}\theta & 0 & 0 \\ \hline 0 & 0 & \text{Cos}\theta & \text{Sin}\theta \\ 0 & 0 & -\text{Sin}\theta & \text{Cos}\theta \end{array}\right] \begin{Bmatrix} u_{1x} \\ u_{1y} \\ \hline u_{2x} \\ u_{2y} \end{Bmatrix}_g^e$$

or

$$\mathbf{u}_\ell^e = \mathbf{T}(\theta)\mathbf{u}_g{}^e. \tag{7.5}$$

The same type of coordinate transformation will apply to components of the element force vector, $\mathbf{P}^e$, namely:

$$\mathbf{P}_\ell^e = \mathbf{T}(\Theta)\mathbf{P}_g{}^e. \tag{7.6}$$

Notice that the transformation matrix is square. Therefore, the inverse transformation can be found by inverting the matrix **T**. Therefore,

$$\mathbf{u}_g = \mathbf{T}^{-1}\mathbf{u}_\ell \text{ , } \mathbf{P}_g = \mathbf{T}^{-1}\mathbf{P}_\ell \text{ .} \tag{7.7}$$

If one carries out the inversion process, an interesting result is obtained. Specifically, we find that the inverse of the transformation is the same as its transpose. This is always true, and it makes our calculations much easier since we can write

$$\mathbf{T}^{-1} = \mathbf{T}^T. \tag{7.8}$$

A matrix with this property is called an **orthogonal** matrix. Therefore, the simple way to write the inverse transformation is

$$\mathbf{u}_g = \mathbf{T}^T\mathbf{u}_\ell. \tag{7.9}$$

7.4 Transformation of Element Matrices

Our ultimate goal is to solve the global equilibrium equations. This requires that all elements be referred to a single global coordinate system and that the assembly of element contributions be relative to that system. Therefore, before we can assemble the element stiffness and load matrices they must be written relative to the global axes.

This means that we need to define global versions of the element matrices, say $\mathbf{S}_g{}^e$ and $\mathbf{C}_g{}^e$. Clearly they are somehow related to the corresponding local element matrices, $\mathbf{S}_\ell^e$ and $\mathbf{C}_\ell^e$. To gain some insight into the relation between the two systems recall that the element behavior was defined in terms of the total potential energy, π^e, of the element. Since that quantity is a scalar, its value must be the same regardless of whether it is computed in element coordinates or global coordinates. If we compute π^e using Eq. (4.67) in local coordinates the result is

$$\pi^e = 1/2\ \mathbf{u}_\ell^{e^T} \mathbf{S}_\ell^e \mathbf{u}_\ell^e - \mathbf{u}_\ell^{e^T} \mathbf{C}_\ell^e. \tag{7.10}$$

By way of comparison, if it is calculated in global coordinates

$$\pi^e = 1/2\ \mathbf{u}_g^{e^T} \mathbf{S}_g^e \mathbf{u}_g^e - \mathbf{u}_g^{e^T} \mathbf{C}_g^e. \tag{7.11}$$

The two forms can be more easily compared if Eq. (7.10) is also written in terms of the global components of the displacements of the element. Before doing that, let us recall the form of the element stiffness and load matrices. From Eqs. (4.64-66) we have

$$\mathbf{S}_\ell^e = \frac{E^e A^e}{\ell^e} \begin{bmatrix} 1 & -1 \\ -1 & 1 \end{bmatrix}$$

and

$$\mathbf{C}_\ell^e = \begin{Bmatrix} C_{1x} \\ C_{2x} \end{Bmatrix}$$

where C_1 and C_2 represent the resultant loads along the local x-axis. Since the global structure will have two displacements per node it will be useful to rewrite the element equations in terms of two local displacements per node. Specifically, the expanded element equations are

$$\frac{E^e A^e}{\ell^2} \left[\begin{array}{cc|cc} 1 & 0 & -1 & 0 \\ 0 & 0 & 0 & 0 \\ \hline -1 & 0 & 1 & 0 \\ 0 & 0 & 0 & 0 \end{array}\right] \left\{\begin{array}{c} u_{1x} \\ u_{1y} \\ \hline u_{2x} \\ u_{2y} \end{array}\right\}_\ell^e = \left\{\begin{array}{c} c_{1x}^e \\ 0 \\ \hline c_{2x}^e \\ 0 \end{array}\right\}_\ell .$$

Note that the stiffness matrix has been expanded by adding rows and columns of zeros to correspond to the local y displacement. That was done because the element can not resist loads in the local y direction. The above expanded element matrices would be substituted into Eq. (7.10). Next substituting the identity of Eq. (7.5) into Eq. (7.10) yields

$$\pi^e = 1/2\ (\mathbf{T}^e \mathbf{u}_g{}^e)^T\ \mathbf{S}_\ell{}^e (\mathbf{T}^e \mathbf{u}_g{}^e) - (\mathbf{T}^e \mathbf{u}_g{}^e)^T \mathbf{C}_g{}^e$$

or

$$\pi^e = 1/2\ \mathbf{u}_g{}^{e^T} (\mathbf{T}^{e^T}\ \mathbf{S}_\ell{}^e \mathbf{T}^e) \mathbf{u}_g{}^e - \mathbf{u}_g{}^{e^T} (\mathbf{T}^{e^T} \mathbf{C}_g{}^e).$$

Comparing this scalar with the same quantity in Eq. (7.11) gives the desired identities

$$\mathbf{S}_g^e = \mathbf{T}^{e^T} \mathbf{S}_\ell{}^e \mathbf{T}^e \tag{7.12}$$

and

$$\mathbf{C}_g{}^e = \mathbf{T}^{e^T} \mathbf{C}_\ell{}^e. \tag{7.13}$$

Of major importance here is that Eqs. (7.12) and (7.13) are not restricted to truss elements. For certain types of elements it would be simpler to form the global element matrices numerically by matrix multiplication.

For the truss element in two dimensions the products in these transformations are easily written out. The results are

$$S^e = \frac{E^e A^e}{\ell^e} \begin{bmatrix} \lambda\lambda & \lambda\mu & -\lambda\lambda & -\lambda\mu \\ \lambda\mu & \mu\mu & -\lambda\mu & -\mu\mu \\ -\lambda\lambda & -\lambda\mu & \lambda\lambda & \lambda\mu \\ -\lambda\mu & -\mu\mu & \lambda\mu & \mu\mu \end{bmatrix}^e \tag{7.14}$$

and

$$C^e = \begin{Bmatrix} C_{1x}\lambda \\ -C_{1x}\mu \\ C_{2x}\lambda \\ -C_{2x}\mu \end{Bmatrix}^e \tag{7.15}$$

where $\lambda = \mathrm{Cos}\,\phi_x$, $\mu = \mathrm{Cos}\,\phi_y = \mathrm{Sin}\,\phi_x$. From Eq. (7.11) an alternate form of these equations is

$$S^e = \frac{A^e E^e}{(\ell^e)^3} \begin{bmatrix} \ell_x^2 & \ell_x \ell_y & -\ell_x^2 & -\ell_x \ell_y \\ \ell_x \ell_y & \ell_y^2 & -\ell_x \ell_y & -\ell_y^2 \\ -\ell_x^2 & -\ell_x \ell_y & \ell_x^2 & \ell_x \ell_y \\ -\ell_x \ell_y & -\ell_y^2 & \ell_x \ell_y & \ell_y^2 \end{bmatrix}^e \qquad (7.16)$$

$$C^e = \frac{1}{\ell^e} \begin{Bmatrix} c_{1x} \ \ell_x \\ -c_{1x} \ \ell_y \\ c_{2x} \ \ell_x \\ -c_{2x} \ \ell_y \end{Bmatrix}^e . \qquad (7.17)$$

A similar set of transformed global stiffness and force vectors can be obtained for a truss element located in three-dimensional space.

7.4 Example Structures

Consider the example structure shown in Fig. 7.2. Assume that all three members have the same area and modulus of elasticity. The structure is described by

Element	ℓ_x	ℓ_y	ℓ	Topology		EA
1	4	3	5	1	2	1000
2	8	0	8	1	3	1000
3	4	-3	5	2	3	1000

The structure is pinned at node 1 and on a horizontal roller at node 3. No distributed loads or thermal loads are considered on the bars. Thus for each element $\mathbf{C}^e = \mathbf{0}$. Only nodal loads are externally applied. Their values, at node 2 are $P_x = 10$, and $P_y = -20$. From Eq. (7.16) the element stiffness matrices, when transformed to the global axes, have the values of

e = 1:

$$S^e = \frac{1000}{125} \begin{bmatrix} 16 & 12 & -16 & -12 \\ 12 & 9 & -12 & -9 \\ -16 & -12 & 16 & 12 \\ -12 & -9 & 12 & 9 \end{bmatrix} \begin{matrix} \text{Global} \\ 1 \\ 2 \\ 3 \\ 4 \end{matrix}$$

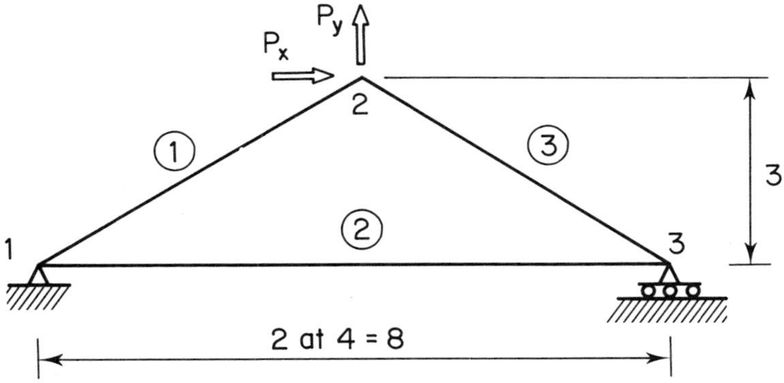

(a) Unsymmetric loading

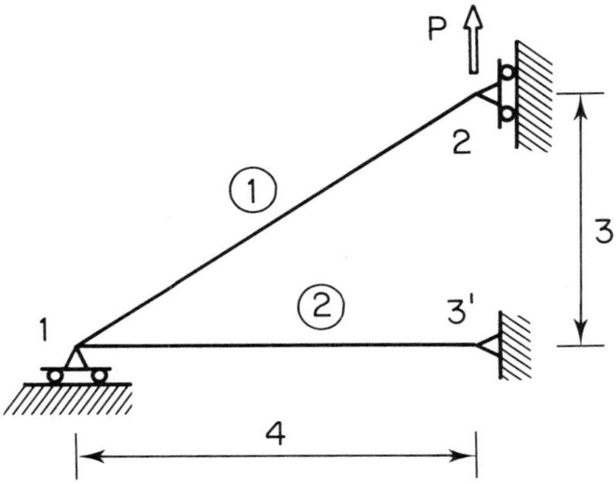

(b) Symmetric loading

Fig. 7.2 A Three Bar Truss Structure

e = 2:

$$S^e = \frac{1000}{512}\begin{bmatrix} 64 & 0 & -64 & 0 \\ 0 & 0 & 0 & 0 \\ -64 & 0 & 64 & 0 \\ 0 & 0 & 0 & 0 \end{bmatrix} \begin{matrix} \text{Global} \\ 1 \\ 2 \\ 5 \\ 6 \end{matrix}$$

e = 3:

$$S^e = \frac{1000}{125}\begin{bmatrix} 16 & -12 & -16 & 12 \\ -12 & 9 & 12 & -9 \\ -16 & 12 & 16 & -12 \\ 12 & -9 & -12 & 9 \end{bmatrix} \begin{matrix} \text{Global} \\ 3 \\ 4 \\ 5 \\ 6 \end{matrix} .$$

The assembled system equilibrium equations are

$$\begin{bmatrix} (128+125) & 96 & -128 & -96 & -125 & 0 \\ & 72 & -96 & -72 & 0 & 0 \\ & & (128+128) & (96-96) & -128 & 0 \\ & & & (72+72) & 96 & -72 \\ & & & & (125+128) & -96 \\ \text{symmetric} & & & & & 72 \end{bmatrix} \begin{Bmatrix} u_1 \\ v_1 \\ u_2 \\ v_2 \\ u_3 \\ v_3 \end{Bmatrix} = \begin{Bmatrix} 0 \\ 0 \\ P_x \\ P_y \\ 0 \\ 0 \end{Bmatrix} .$$

However, three displacements are prescribed to be zero. Modifying the above equations to include the boundary conditions reduces them to

$$\begin{bmatrix} 256 & 0 & -128 \\ & 144 & 96 \\ \text{Sym.} & & 253 \end{bmatrix} \begin{Bmatrix} u_2 \\ v_2 \\ u_3 \end{Bmatrix} = \begin{Bmatrix} P_x \\ P_y \\ 0 \end{Bmatrix} .$$

These equations can be inverted by hand or by using I3BY3, or SYMINV. The result is

$$\begin{Bmatrix} u_2 \\ v_2 \\ u_3 \end{Bmatrix} = \frac{1}{4.608 \times 10^6}\begin{bmatrix} 27216 & & \text{Sym.} \\ -12288 & 48384 & \\ 18432 & -24576 & 36864 \end{bmatrix} \begin{Bmatrix} P_x \\ P_y \\ 0 \end{Bmatrix} .$$

Substituting the given load values yields

$$[u_2 \; v_2 \; u_3] = [0.1124 \quad -0.2367 \quad 0.1467].$$

Note that if P_x had been zero then $u_2 = u_3/2 = 0.0533$, and $v_2 = -0.21$. That is, the deformation of the structure would have been symmetric with respect to the center of the truss.

The concepts of symmetry and anti-symmetry are often useful in finite element analysis. It is common to find half, quarter, or eighth order symmetry conditions that can reduce the analysis cost to the square of the corresponding fractional part of a total analysis cost. Half symmetry was employed in Sec. 4.6. Here we will apply it to the above truss. First we view the loads, members, and supports as viewed relative to a mirror placed at the symmetry section. The result is shown in Fig. 7.2b. The applied loads and the stiffness of members lying in the symmetry plane are reduced by half. The nodes or member midpoints that lie in the symmetry plane are allowed to move only in that plane. Any supports that are not on the symmetry plane can be modified to support the structure in a consistent manner when viewed from the symmetry plane. This means that our simplified structure can be described as

Element	ℓ_x	ℓ_y	ℓ	Topology		EA
1	4	3	5	1	2	1000
2	4	0	4	1	3	1000 .

The stiffness for the third element is no longer needed. The first member is unchanged. The length of the second member is cut in half so its stiffness doubles. The assembled elements give

$$\begin{bmatrix} (128+250) & 96 & -128 & -96 & -250 & 0 \\ & 72 & -96 & -72 & 0 & 0 \\ & & 128 & 96 & 0 & 0 \\ & & & 72 & 0 & 0 \\ & & & & 250 & 0 \\ \text{symmetric} & & & & & 0 \end{bmatrix} \begin{Bmatrix} u_1 \\ v_1 \\ u_2 \\ v_2 \\ u_{3'} \\ v_{3'} \end{Bmatrix} = \begin{Bmatrix} 0 \\ 0 \\ 0 \\ P_y/2 \\ 0 \\ 0 \end{Bmatrix} .$$

Points in the plane of symmetry must always move in that plane. Thus $u_1 = u_{3'} = 0$. Conversely node 1 must be able to move normal to the plane of symmetry. Thus $u_1 \neq 0$. Node 2 has an external load, P, applied tangent to the plane of symmetry. Thus it must be allowed to move tangent to the plane ($v_2 \neq 0$) so that the force can do work on the structure. That is, in a given direction one can specify either the force or the displacement at a point, but not both. Clearly node 1 has an unknown reaction that is parallel to the symmetry plane. Thus it must be restrained in that direction, $v_1 = 0$. The restrained structural stiffness is

$$\begin{bmatrix} 378 & -96 & 0 \\ & 72 & 0 \\ \text{sym.} & & 0 \end{bmatrix} \begin{Bmatrix} u_1 \\ v_2 \\ v_{3'} \end{Bmatrix} = \begin{Bmatrix} 0 \\ -20/2 \\ 0 \end{Bmatrix} .$$

However, these equations are still singular after the application of the usually symmetric conditions. Note that the third row and column are zero. This means that there is no stiffness associated with the displacement $v_{3'}$. From the original structure in Fig. 7.2a we note that the center of member 2 must have a zero vertical deflection. Employing this additional physical insight we can now also state that $v_{3'} = 0$. Therefore, for a symmetric structure with symmetric loads the equilibrium equations, relative to the plane of symmetry are

$$\begin{bmatrix} 378 & -96 \\ -96 & 72 \end{bmatrix} \begin{Bmatrix} u_1 \\ v_2 \end{Bmatrix} = \begin{Bmatrix} 0 \\ -10 \end{Bmatrix}$$

so the results are

$$\begin{Bmatrix} u_1 \\ v_2 \end{Bmatrix} = \begin{Bmatrix} -0.05333 \\ -0.21 \end{Bmatrix}$$

as before except for the sign change on u_1. This solution shows that for the above example there are only two degrees of freedom required when symmetry is available versus the three that were used before. For this simple example there was not much difference in the computational effort required in the symmetric and non-symmetric solutions. However, if there are hundreds or thousands of symmetric elements then the cost saving is very significant when a symmetric analysis can be utilized.

8. BEAM ANALYSIS

8.1 Introduction

A common structural system considered in engineering is that of the elastic beam. Such a beam is shown in Fig. 8.1. In mechanics of materials a number of common assumptions are made in order to reduce the analysis to a one-dimensional formulation. The most common assumption is that planes in the beam, normal to a fiber along the x-axis, remain normal to that fiber in its deformed state. This assumption makes the axial displacement, u, and the axial strain, ε, vary linearly with the transverse coordinate, y. Let v denote the transverse displacement, and $\theta = v'$ is the slope of the beam. Then the axial displacement relation, for small slopes, is

$$u(x,y) = -yv' = -y\frac{dv}{dx} . \tag{8.1}$$

The axial strain is

$$\varepsilon(x,y) = \frac{du}{dx} = -y\frac{d^2v}{dx^2} = -yv''. \tag{8.2}$$

For an elastic material the stress, σ, is defined by Hooke's law as

$$\sigma(x,y) = E(x)\ \varepsilon(x,y) \tag{8.3}$$

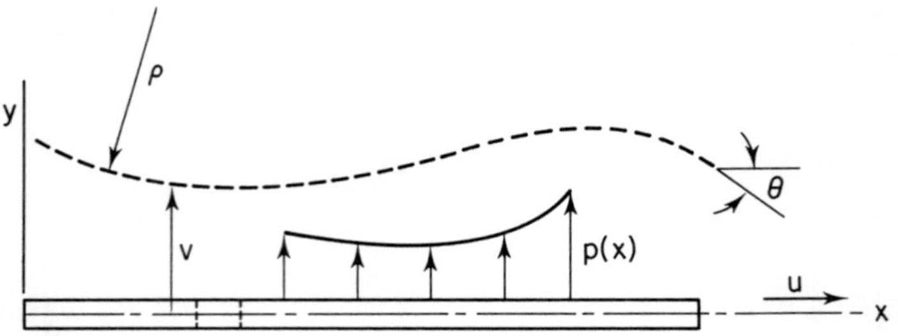

(a) Coordinates and notation

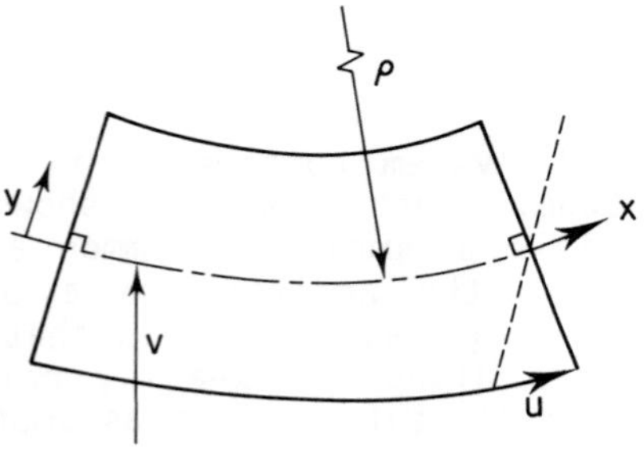

(b) Displacements when plane sections remain plane

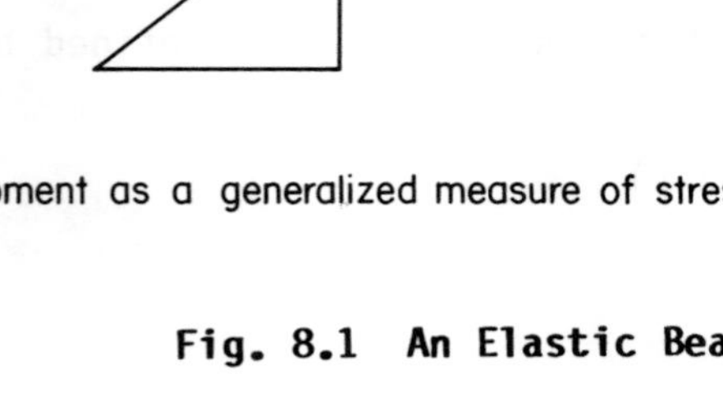

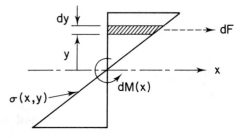

(c) Moment as a generalized measure of stress

Fig. 8.1 An Elastic Beam

where E is the elastic modulus of the material. These quantities could vary with both of the spatial coordinates. We desire to formulate a one-dimensional model. We will define a **generalized strain** and a **generalized stress** to accomplish this goal.

From calculus we should recall that the quantity $v''(x) = 1/\rho$ is known as the **curvature** of the deflected beam and ρ is the radius of curvature. The signs in Eqs. (8.1) and (8.2) have been chosen so that a positive sign denotes tension. From statics one can show that the resultant axial load from the distributed stress is zero. However, there is a non-zero resultant moment. Its value is given by

$$\begin{aligned} M(x) &= \int dm = \int -y\, dF \\ &= \int -y\sigma\, da = \int -yE\varepsilon\, da \\ &= {}_A\!\int Ey^2 v''\, da = EIv''(x) \end{aligned} \qquad (8.4)$$

where A is the cross-sectional area of the member and

$$I(x) = {}_A\!\int y^2 da$$

is the moment of inertia of the cross sectional area. We will call this moment our generalized stress measure since it only depends on x.

In mechanics of materials the deflections of the beam are determined by solving the differential equation of equilibrium:

$$\frac{d^2}{dx^2}\left(EI\frac{d^2v}{dx^2}\right) = p(x) \qquad (8.5)$$

where p(x) is the distributed transverse load per unit length. The four constants of integration are determined by satisfying the boundary conditions on the deflections, v, and slopes, θ. However, in our present study we need an integral formulation for our finite element model.

8.2 Variational Procedure

A variational formulation for the elastic beam is related to minimizing the total energy and work in the system. One important term required for the analysis is the strain energy. That quantity is defined as half the volume integral of the product of the stress and strain. Here we wish to reduce this quantity to a function of x alone. The scalar strain energy is

$$U = 1/2 \int_V \sigma \, \varepsilon \, dV \tag{8.6}$$

or

$$U = 1/2 \int_L \int_{A(x)} (Eyv'')(yv'') \, da \, dx.$$

Since only v" depends on x this reduces to

$$U = 1/2 \int_L E(x) \, I(x)(v'')^2 dx \tag{8.7}$$

or

$$U = 1/2 \int_L M(x) v''(x) dx. \tag{8.8}$$

8.2.1 Mathematical Summary

Comparing Eqs. (8.6) and (8.8) suggests that we should select the curvature, v", as our generalized strain measure. Having made this choice we can use Eq. (8.4) to define a generalized constitutive relation. Defining

$$\boldsymbol{\sigma} = M(x)$$

$$\boldsymbol{\varepsilon} = v''(x) \tag{8.9}$$

and writing Eq. (8.4) as

$$\boldsymbol{\sigma} = \mathbf{D}\boldsymbol{\varepsilon} \tag{8.10}$$

defines

$$\mathbf{D} = E(x) \, I(x) \tag{8.11}$$

as the generalized Hooke's law. The left hand side of the last three equations have been defined as arrays even though they only contain a single term. This is done to give some insight into what would happen for a plate or shell where there would be three curvatures of the surface, three corresponding moments, and **D** would become a 3x3 array involving the material properties and thickness of the section. If we consider a beam of width b and thickness h then Eq. (8.11) could be written as

$$\mathbf{D} = \frac{Eh^3b}{12} .$$

For a plate the generalized stresses and strains would be

$$\boldsymbol{\sigma}^T = [M_{xx} \quad M_{yy} \quad M_{xy}]$$

$$\boldsymbol{\varepsilon}^T = [w_{,xx} \quad w_{,yy} \quad w_{,xy}]$$

where w is the transverse displacement of the plate. The moments are written on a unit length basis (b = 1) so for the plate

$$\mathbf{D} = \frac{Eh^3}{12(1 - \nu^2)} \begin{bmatrix} 1 & 0 & 0 \\ 0 & 1 & 0 \\ 0 & 0 & (1 - \nu)/2 \end{bmatrix} .$$

While we do not plan to consider plates here we will use the generalized symbolism to give insight to such problems.

Equation (8.8) shows that our selections for generalized stress-strain measures will correctly define the strain energy in the system. Next we need to define the work done by the applied loads, P_i, or couples, C_i. The work done by a transverse force is the product of the force and the transverse displacement. Likewise the work done by a couple is the product of the couple and rotation (slope) at its point of application. These contributions define a work term, W, given by

$$W = \int_L v(x)p(x)dx + \sum_i v(x_i)P_i + \sum_j v'(x_j)C_j .$$

The last two terms represent work done by concentrated point loads or couples. Thus the total potential energy, $\pi = U - W$ is

$$\pi = 1/2 \int_L EI(v''(x))^2 dx - \int_L v(x)p(x)dx$$
$$- \sum_i v_i P_i - \sum_j v'_j C_j . \qquad (8.12)$$

To determine the displacement field, v(x), that corresponds to the equilibrium state we must minimize π and satisfy the boundary conditions on v and $v' = \theta$.

8.2.2 Element Matrices

To introduce our finite elements we select a series of line segments to make up the region, L. There are numerous elements that could be selected. First we will select an element with two nodes. Next it is necessary to assume a displacement approximation so we can evaluate the potential energy in Eq. (8.10). That equation contains second derivatives and thus we need to assume a solution for v that will have both the deflection, v, and the slope, v', continuous between elements. The most common assumption is to select the cubic Hermite polynomial presented in Fig. 5.5. The unknowns at each of the two element nodes are v and $v' = \theta$. These quantities will be called our **generalized displacements** or the generalized degrees of freedom.

Thus our element interpolation functions are

$$v(x) = [H_1^e(x)\ H_2^e(x)\ H_3^e(x)\ H_4^e(x)] \begin{Bmatrix} v_1 \\ v_1' \\ v_2 \\ v_2' \end{Bmatrix}^e$$

or

$$v(x) = \mathbf{H}^e(x)\mathbf{a}^e \tag{8.13}$$

where $\mathbf{a}^e$ denotes the generalized displacements of the element. The derivatives of the displacements are

$$v'(x) = \theta(x) = \mathbf{H}^{e'}(x)\mathbf{a}^e$$

and

$$v''(x) = \mathbf{H}^{e''}(x)\ \mathbf{a}^e. \tag{8.14}$$

Since v" and $\mathbf{a}^e$ have been selected as our generalized strains and generalized displacements we will use the notation of Eq. (4.60) and write Eq. (8.12) as

$$\boldsymbol{\varepsilon}^e = \mathbf{B}^e \mathbf{a}^e \tag{8.15}$$

where $\varepsilon = v''$ in our present study. In the study of plates and shells additional curvature terms would be present in ε.

Employing our generalized notation the stiffness matrix and distributed load vector can be written by inspection as

$$\mathbf{K}^e = \int_{\ell^e} \mathbf{B}^e(x)^T \mathbf{D}^e(x) \mathbf{B}^e(x) dx \tag{8.16}$$

$$F_p^{\ e} = \int_{\ell^e} \mathbf{H}^e(x)^T p^e(x) dx. \tag{8.17}$$

Here we will again use unit coordinates on the element and set $r = x/\ell$ so $(\ )' = d()/dx = d()/dr*dr/dx = d()/dr*\ 1/\ell$. Thus

$$\mathbf{B}^e = \mathbf{H}^{e''} = \frac{1}{\ell^2}\frac{d^2\mathbf{H}}{dr^2}$$

$$= \frac{1}{\ell^2}\left[(12r-6) \quad \ell(6r-4) \quad (6-12r) \quad \ell(6r-2)\right].$$

Recalling Eq. (6.5),

$$\int_{\ell} r^m dx = \ell/(m+1)$$

and assuming $\mathbf{D}^e$ is a constant then the stiffness is

$$\mathbf{K}^e = \frac{EI}{\ell^3}\begin{bmatrix} 12 & & & \text{sym.} \\ 6\ell & 4\ell^2 & & \\ -12 & -6\ell & 12 & \\ 6\ell & 2\ell^2 & -6\ell & 4\ell^2 \end{bmatrix}. \tag{8.18}$$

If the lateral load is constant then

$$F_p^{\ e} = p^e \int_{\ell^e} \mathbf{H}^{e^T} dx = p^e \ell^e \int_0^1 \mathbf{H}^{e^T}(r) dr$$

or simply

$$\mathbf{F}_p^e = p^e \ell^e \begin{Bmatrix} 1/2 \\ \ell^e/12 \\ 1/2 \\ -\ell^e/12 \end{Bmatrix}. \tag{8.19}$$

This result for the consistent nodal loads is illustrated in Fig. 8.2. Note that the distributed load puts half the resultant load at each end. It also causes a nodal couple at each of the two nodes.

When we wrote Eq. (8.12) we assumed that point loads would only be applied at the node points. This may not always be true and we should consider such a load condition as a special case of a distributed load. In that case the length of the distributed load approaches zero and the magnitude of the force per unit length approaches infinity, but the resultant load P is constant. That is, we define the load to be

$$p(x) = P\,\delta(x - x_0)$$

where δ is the Dirac Delta. Then the generalized load vector is

$$\mathbf{F}_p^e = \int_{\ell^e} \mathbf{H}^{e^T}(x) P\,\delta(x - x_0)\,dx$$

which is integrated by inspection by using the integral property of δ in Eq. (4.7) to yield

$$\mathbf{F}_p^e = P\mathbf{H}^e(r_0)$$

where $r_0 = x_0/\ell$ is the point of application of the load. This result is also shown in Fig. 8.2. To check this concept, assume that the load is at node 1. Then $r_0 = 0$ and Fig. 8.2 gives

$$\mathbf{F}_p^{e^T} = P[1 \quad 0 \quad 0 \quad 0]$$

as expected. That is, all the force goes into node 1 and no element nodal moments are generated.

Other common loading conditions can be treated in the same way. For example, if p(x) varies linearly from p_1^e to p_2^e at the nodes of element e then

$$p(x) = (1 - x/\ell)p_1^{\,e} + x/\ell\; p_2^{\,e}$$
$$= [(1 - r) \quad r]\begin{Bmatrix} p_1 \\ p_2 \end{Bmatrix}^e \tag{8.20}$$

and

$$\mathbf{F}_p^{\,e} = \int_{\ell^e} \begin{Bmatrix} 1 - 3r^2 + 2r^3 \\ \ell(r - 2r^2 + r^3) \\ 3r^2 - 2r^3 \\ \ell(r^3 - r^2) \end{Bmatrix} [(1 - r) \quad r]\, dx \begin{Bmatrix} p_1 \\ p_2 \end{Bmatrix}^e$$

$$= \frac{\ell^e}{20} \begin{bmatrix} 7 & 3 \\ \ell & 2\ell/3 \\ 3 & 7 \\ -2\ell/3 & -\ell \end{bmatrix} \begin{Bmatrix} p_1 \\ p_2 \end{Bmatrix}^e \quad . \tag{8.21}$$

If the load is constant so $p_1^{\,e} = p_2^{\,e} = p$ then this reduces to Eq. (8.19), as expected. Likewise if $p_1^{\,e} = 0$ and $p_2^{\,e} = p$ this defines a triangular load and

$$\mathbf{F}_p^{\,e^T} = \frac{p\ell}{20} [3 \quad 2\ell/3 \quad 7 \quad -\ell] \quad . \tag{8.22}$$

It is common to tabulate such results in terms of an applied unit resultant load. That resultant is

$$R^e = \int_{\ell^e} p^e(x)dx \quad .$$

For common load variations, such as constant, linear, parabolic, and cubic the resultant loads are, respectively,

$$R^e = p\ell$$
$$R^e = p\ell/2$$
$$R^e = p\ell/3$$
$$R^e = p\ell/4 \; .$$

The location, $\bar{x}$, of the resultant applied load is found from

$$\bar{x}R^e = \int_{\ell^e} px\ dx$$

and the corresponding results are $\bar{x} = \ell/2$, $\bar{x} = 2\ell/3$, $\bar{x} = 3\ell/4$, and $\bar{x} = 4\ell/5$. Thus if we normalize Eq. (8.22) by dividing by the resultant load, $p\ell/2$, the result is

$$\mathbf{f}_p^{e^T} = [3/10 \quad \ell/15 \quad 7/10 \quad -\ell/10]$$

which is the result in Fig. 8.2. We can also check the unit load results by applying statics to the data in Fig. 8.2. To check the triangular load summary we first take the sum of the moments about node 1. This gives

$$\begin{aligned} +(2\ell/3)1 &= 0 + (7/10)\ell + \ell/15 - \ell/10 \\ 2\ell/3 &= \ell(21 + 2 - 3)/30 \\ 2\ell/3 &= 2\ell/3, \text{ ok.} \end{aligned}$$

Similarly the sum of the moments about node 2 is

$$\begin{aligned} -(\ell/3)1 &= -\ell(3/10) + 0 + \ell/15 - \ell/10 \\ -\ell/3 &= \ell(-9 + 2 - 3)/30 \\ -\ell/3 &= -\ell/3, \text{ ok.} \end{aligned}$$

8.3 Sample Application

To present an analytic example of this element consider a single element solution of the cantilever beam shown in Fig. 8.3 to determine the deflection and slope at the free end. Usually the deflected shape of a beam is defined by a fourth or fifth order polynomial in x. Thus our cubic element solution will usually give only an approximate solution. For a single element the system equations are

$$\frac{EI}{L^3}\begin{bmatrix} 12 & 6L & -12 & 6L \\ 6L & 4L^2 & -6L & 2L^2 \\ \hline -12 & -6L & 12 & -6L \\ 6L & 2L^2 & -6L & 4L^2 \end{bmatrix}\begin{Bmatrix} v_1 \\ \theta_1 \\ \hline v_2 \\ \theta_2 \end{Bmatrix} = \frac{WL}{2}\begin{Bmatrix} 3/10 \\ L/15 \\ 7/10 \\ -L/10 \end{Bmatrix}. \tag{8.23}$$

The right side support requires that $v_2 = 0 = \theta_2$. The reduced equations become

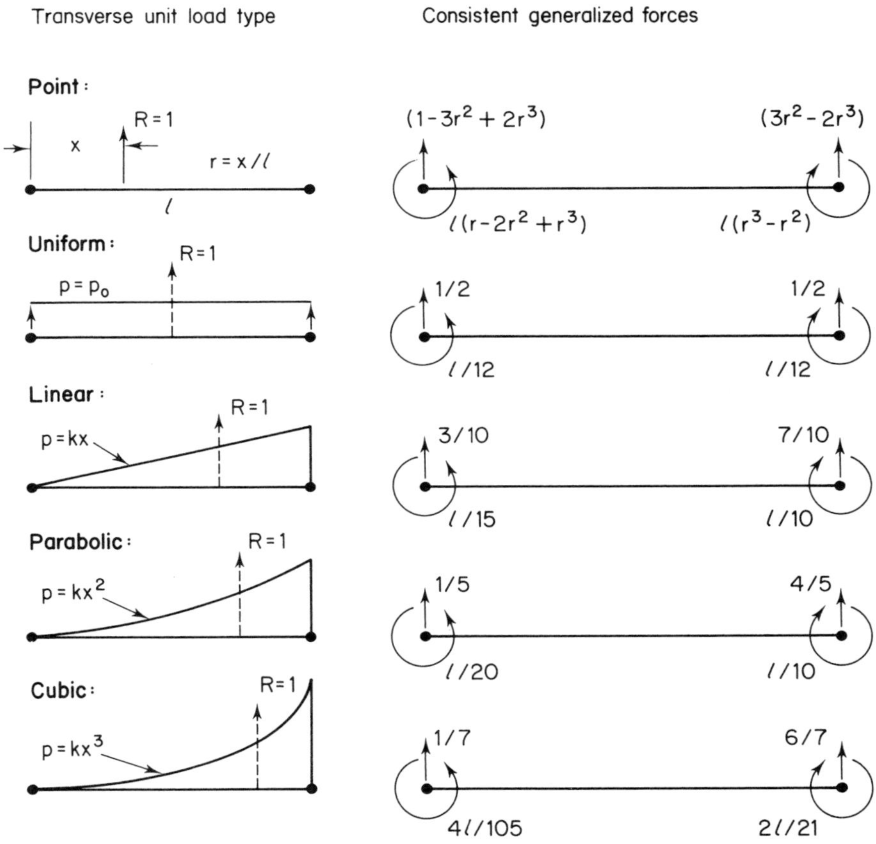

Fig 8.2 Consistent Nodal Loads and Couples Due to Distributed Element Load

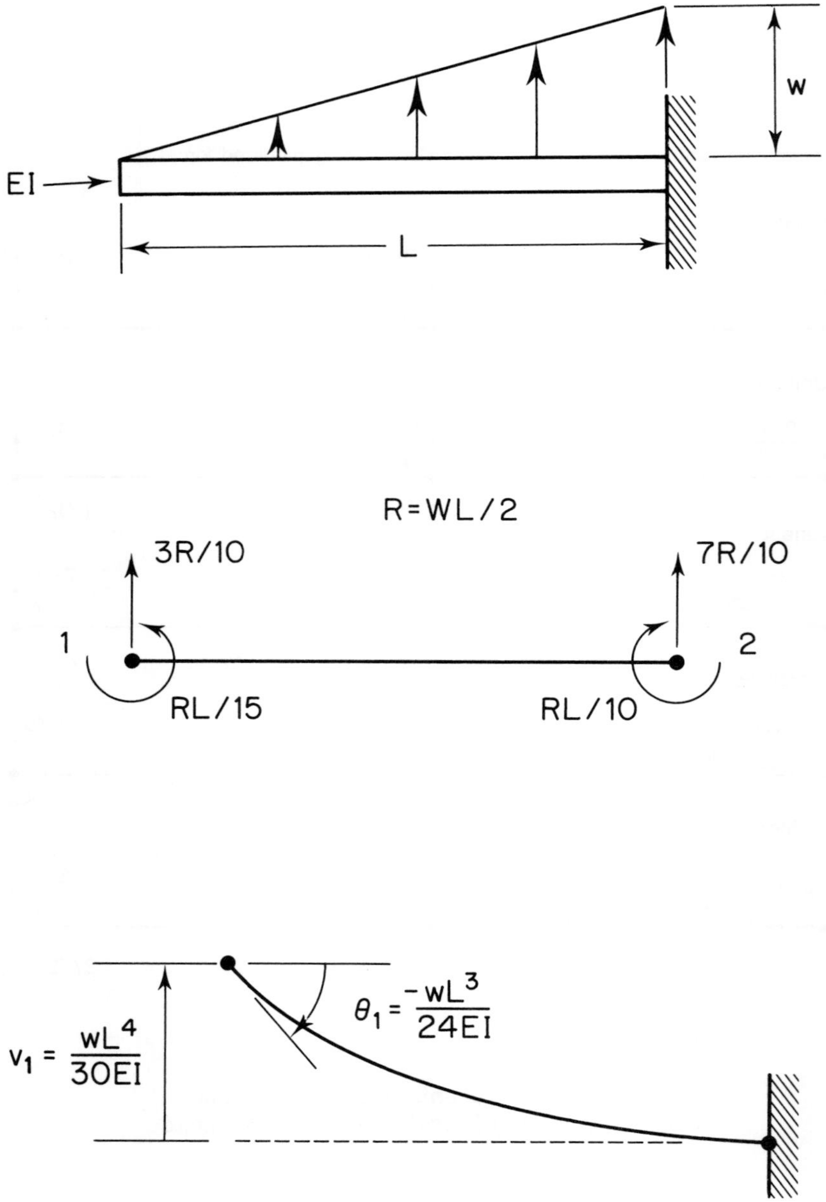

Fig. 8.3 A Single Element Approximate Solution

$$\frac{EI}{L^3}\begin{bmatrix} 12 & 6L \\ 6L & 4L^2 \end{bmatrix}\begin{Bmatrix} v_1 \\ \theta_1 \end{Bmatrix} = \frac{WL}{2}\begin{Bmatrix} 3/10 \\ L/15 \end{Bmatrix}$$

so

$$\begin{Bmatrix} v_1 \\ \theta_1 \end{Bmatrix} = \frac{L^3}{12\ EIL^2}\begin{bmatrix} 4L^2 & -6L \\ -6L & 12 \end{bmatrix}\begin{Bmatrix} 3/10 \\ L/15 \end{Bmatrix}\frac{WL}{2}$$

$$\begin{Bmatrix} v_1 \\ \theta_1 \end{Bmatrix} = \frac{WL^3}{EI}\begin{Bmatrix} L/30 \\ -1/24 \end{Bmatrix}. \tag{8.24}$$

The exact solution is $120\ EI\ v = wL^4(4 - 5x/L + (x/L)^5)$ so the exact values of the maximum deflection and slope are $v = WL^4/(30EI)$ and $\theta = -WL^3/(24EI)$, respectively. Thus our single element solution gives the exact values of both v and θ at the nodes, but is only approximate in the interior of the element.

It is desirable to correlate the deflection, v, and its derivatives with the physical quantities that they represent in the beam. They are

Deflection	$= v$	
Slope	$= v' = dv/dx = \theta$	
Moment	$= EI\ v''$	
Shear	$= dM/dx = EI\ v'''$	(EI constant)
Load	$= dV/dx = EI\ v''''$	(EI constant).

Thus quantities like moment and shear are represented by higher derivatives. They are often used to design a member. Thus an important question is, where are the moments and shears most accurate? Or, where are v''' and v'''' most accurate? This topic is considered in Chap. 10.

8.4 Element Equations via Galerkin Method

Here we will illustrate the development of the element matrices by applying Galerkin's Method to the governing differential equation. Recall that for a beam subject to a load p(x) the differential equation describing the elastic curve is given by Eq. (8.5)

$$\frac{d^2}{dx^2}\left(EI\frac{d^2v}{dx^2}\right) - p(x) = 0.$$

If we substitute an approximate solution, u(x), then this gives

$$\frac{d^2}{dx^2}\left(EI\frac{d^2u}{dx^2}\right) - p(x) = R(x)$$

where R(x) is the residual error term. Interpolating the approximate solution with Eq. (8.13) gives

$$u(x) = \mathbf{H}^e(x)\,\mathbf{a}^e = \mathbf{a}^{e^T}\mathbf{H}^{e^T}.$$

Applying Galerkin's criterion to the error term gives

$$0 \equiv \int u(x)\,R(x)\,dx$$

$$= \sum_e \int_{L^e} u(x)\,R(x)\,dx$$

$$= \sum_e \int_{L^e} \mathbf{a}^{e^T}\mathbf{H}^{e^T}(x)R(x)dx.$$

But the array $\mathbf{a}^e$ is a vector of arbitrary constants. This implies that we require

$$\mathbf{0} = \int_{L^e} \mathbf{H}^{e^T}(x)R(x)dx$$

or

$$\mathbf{0} = \int_{L^e} \mathbf{H}^{e^T}\left[\frac{d^2}{dx^2}\,EI\frac{d^2u}{dx^2} - p(x)\right]dx$$

as suggested in Eq. (4.19). Let a prime denote a derivative. Then twice integrating the first term by parts gives

$$\int EI\mathbf{H}''^T u''dx + EI\,\mathbf{H}^T u'''\Big|_0^L - EI\,\mathbf{H}'^T u''\Big|_0^L - \int \mathbf{H}^T\,pdx = \mathbf{0}.$$

Substituting the interpolation for u the first integral gives

$$\int_{L^e} EI\ H''^T(x)\mathbf{H}''^T(x)dx\ \mathbf{a}^e = \mathbf{S}^e\mathbf{a}^e$$

where

$$\mathbf{S}^e = \int_{L^e} EI\mathbf{H}''^T\mathbf{H}''\ dx$$

is the element stiffness matrix given earlier in Eq. (8.16). The second integral is the consistent force vector given earlier in Eq. (8.17). The remaining two terms define the natural boundary conditions on the beam.

8.5 **Exercises**

1. Use statics to check the parabolic and cubic load summaries shown in Fig. 8.2.
2. Use Eq. (8.24) to compute the reactions in Eq. (8.23) and compare them with the exact results from statics.
3. Compare the exact end deflection and slope of the cantilever with those from a single beam element solution. Replace the load vector in Eq. (8.23) with that due to (a) a point load at the end (b) a uniform load.
4. Determine the reactions in problem (3) above.
5. Resolve Eq. (8.23) for a propped cantilever beam with a triangular load. That is, find θ_1, if $v_1 = 0$ also.
6. Obtain a single element approximation for a simply supported beam with a concentrated center load. Use (a) the full beam (b) a half symmetry model.
7. Utilize Eq. (8.23) to solve the problem where the cantilever beam is supported at node 1 instead of node 2.

9 Transient Analysis, One-Dimensional Examples*

9.1 Introduction

Many problems require the solution of time-dependent equations. In this context, there are numerous theoretical topics that an analyst should investigate before selecting a computational algorithm. These include the stability limits, amplitude error, phase error, etc., [10], [16], [24]. The purpose here is to cite typical procedures that arise in time integration problems. Only transient problems with first order time derivatives will be considered here.

Recall the calculus definition of the derivative of a quantity, R, with respect to time:

$$\dot{R} = \frac{dR}{dt} = \lim_{\Delta t \to 0} \frac{R(t + \Delta t) - R(t)}{\Delta t} .$$

We are not going to employ the theory of differential equations. Thus we will make an approximation to replace the above temporal operator with an algebraic approximation. The most commonly used finite difference approximations are the forward difference

$$\dot{R} \simeq (R(t + \Delta t) - R(t))/\Delta t,$$

the backward difference,

$$\dot{R} = (R(t) - R(t - \Delta t))/\Delta t,$$

and central difference

$$\dot{R} = (R(t + \Delta t) - R(t - \Delta t))/(2\Delta t),$$

where Δt denotes a small time difference. Similar expressions can be developed to estimate the second derivative. The actual time integration algorithm is determined by the choices of the difference operators and the ways that they are combined. These expressions are called finite differences since the time difference Δt is finite instead of the infinitesimal limit in the derivative.

9.2 Time Integration Approximations

The governing system of equations for a transient application are generally ordinary differential equations of the form

$$\mathbf{A}\,\dot{\mathbf{R}}(t) + \mathbf{B}\,\mathbf{R}(t) = \mathbf{P}(t) \qquad (9.1)$$

where $(\dot{\ })$ = d( )/dt denotes the derivative with respect to time. Generally, one or more of the coefficients in **R**, say R_j, will be defined in a boundary condition as a function of time, i.e. $R_j = g(t)$. Also, the initial values, **R**(0), must be known to start the transient solution. Note that the governing equations now involve two square matrices, **A** and **B**, at the system level. Thus in general it will be necessary to apply the previously discussed square matrix assembly procedure twice. This is,

$$\mathbf{A} = \sum_{e=1}^{NE} \mathbf{A}^e \qquad (9.2)$$

$$\mathbf{B} = \sum_{e=1}^{NE} \mathbf{B}^e$$

where $\mathbf{A}^e$ and $\mathbf{B}^e$ are generated from the direct assembly procedure and the corresponding element contributions, say $\mathbf{a}^e$ and b^e.

Only the direct step-by-step time integration of Eq. (9.1) will be considered. There are many such procedures published in the literature. The text by Myers [16] examines in detail many of the aspects of simple time integration procedures for linear transient applications. He utilizes both finite difference and finite element spatial approximations and illustrates how their transient solutions differ.

9.2.1 The Euler Integration Method

The accuracy, stability, and relative computational cost of a transient integration scheme depend on how one approximates the 'velocity', during the time step. For example, one could assume that the velocity during the time step is (a) constant, (b) equal to the average value at the beginning and end of the step, (c) varies linearly during the step, etc. To illustrate cases (a) and (c) consider a time interval of $k = \Delta t$ and assume a Taylor series for $\mathbf{R}(t)$ in terms of the value at the previous time step, $\mathbf{R}(t - k)$:

$$\mathbf{R}(t) = \mathbf{R}(t - k) + k\dot{\mathbf{R}}(t - k) + k^2/2\ddot{\mathbf{R}}(t - k) + \ldots \qquad (9.3)$$

Then as illustrated in Fig. 9.1, the first assumption gives and the above equation yields $\mathbf{R}(t)$. The standard Euler integration procedure is obtained by a zero acceleration multiplying both sides of Eq. (9.3) by $\mathbf{A}$:

$$\mathbf{A}\ \mathbf{R}(t) = \mathbf{A}\ \mathbf{R}(t - k) + k\mathbf{A}\ \dot{\mathbf{R}}(t - k) \qquad (9.4)$$

and substituting Eq. (9.1) at time $(t - k)$

$$\mathbf{A}\ \dot{\mathbf{R}}(t - k) = \mathbf{P}(t - k) - \mathbf{B}\ \mathbf{R}(t - k)$$

to obtain the final result that

$$\mathbf{A}\ \mathbf{R}(t) = k\mathbf{P}(t - k) + \{\mathbf{A} - k\mathbf{B}\}\mathbf{R}(t - k). \qquad (9.5)$$

One can make the general observation that the governing ordinary differential equations have been reduced to a new set of algebraic equations of the form

$$\mathbf{S}\ \mathbf{R}(t) = \mathbf{F}(t) \qquad (9.6)$$

which must be solved at each time step. In the present case of the Euler method one has system matrices

$$\mathbf{S} = \mathbf{A}$$

and

$$\mathbf{F}(t) = k\mathbf{P}(t - k) + (\mathbf{A} - k\mathbf{B})\mathbf{R}(t - k). \tag{9.7}$$

As illustrated in Table IX.I, all integrations can be reduced to the form of Eq. (9.6). When the problem is linear and the time step, k, is held constant, the system square matrix does not change with time. Thus it need be assembled and 'inverted' only once. Then at each time step it is only necessary to evaluate F(t) and solve for R(t).

Before considering the practical significance of the alternate forms of Eq. (9.6), let us return to the assumption that is linear during the time step. From Fig. 9.1 one notes that

$$\dot{\mathbf{R}}(t) = \dot{\mathbf{R}}(t - k) + k\ddot{\mathbf{R}}(t - k). \tag{9.8}$$

Solving Eq. (9.3) for $\ddot{\mathbf{R}}$ and substituting into the above equation leads to

$$\dot{\mathbf{R}}(t) = 2[\mathbf{R}(t) - \mathbf{R}(t - k)]/k - \dot{\mathbf{R}}(t - k). \tag{9.9}$$

Substituting into Eq. (9.1) at time t yields the system equations

$$\mathbf{S}\ \mathbf{R}(t) = \mathbf{F}(t),$$

where now

$$\mathbf{S} = \mathbf{B} + 2\mathbf{A}/k \tag{9.10}$$

$$\mathbf{F}(t) = \mathbf{P}(t) + \mathbf{A}(2\mathbf{R}(t - k)/k + \dot{\mathbf{R}}(t - k)).$$

This is referred to as the linear velocity algorithm. A comparison of Eqs. (9.7) and (9.10) is useful. The Euler method is known as an explicit method while the linear velocity formulation is one of many implicit algorithms. Note that the Euler form requires no additional storage while the linear velocity algorithm must store the velocity at time t and (t - k), perform the calculations necessary to update its value at each time step. The necessary recurrence relation which utilizes the above calculated values for R(t) is obtained from Eq. (9.9). Also note that the implicit procedure requires one to have initial starting values for the initial velocity. These can be obtained from Eq. (9.1) as

$$\dot{\mathbf{R}}(0) = \mathbf{A}^{-1}(\mathbf{P}(0) - \mathbf{B}\ \mathbf{R}(0)). \tag{9.11}$$

However, this is a practical approach only so long as **A** is a diagonal matrix.

Table IX.I System Matrices for Linear Transients

1. **Euler (forward difference), $k = \Delta t$**

   $$S = A/k$$

   $$F = P(t - k) + (A/k - B)R(t - k)$$

2. **Crank-Nicolson (mid-difference), $h = \Delta t/2$,**

   $$S = A/k + B/2$$

   $$F = P(t - h) + (A/k - B/2)R(t - k)$$

3. **Linear velocity**

   $$S = 2A/k + B$$

   $$F = P(t) + A(2R(t - k)/k - R(t - k) \text{ and}$$

   $$\dot{R}(t) = 2(R(t) - R(t - k))/k + \dot{R}(t - k)$$

9.2.2 Diagonal Matrices for Transient Solutions

When one uses a finite difference spatial formulation, the system matrix **A** is a diagonal matrix. However, if one utilizes a consistent finite element formulation it is not a diagonal matrix. Thus, the consistent finite element form introduces an implied coupling of some of the coefficients in the time derivative matrix. As shown by Myers [16], this tends to result in a less stable time integration algorithm.

Clearly, converting **A** to a diagonal matrix would also save storage and make the evaluation of equations such as Eq. (9.11) much more economical. Some engineering approaches for modifying **A** have been shown to be successful. To illustrate these, consider the form of a typical element contribution, say $\mathbf{a}^e$. The consistent definition for constant properties is

$$\mathbf{a}^e \equiv q \int_{V^e} \mathbf{H}^{e^T} \mathbf{H}^e dv, \qquad (9.12)$$

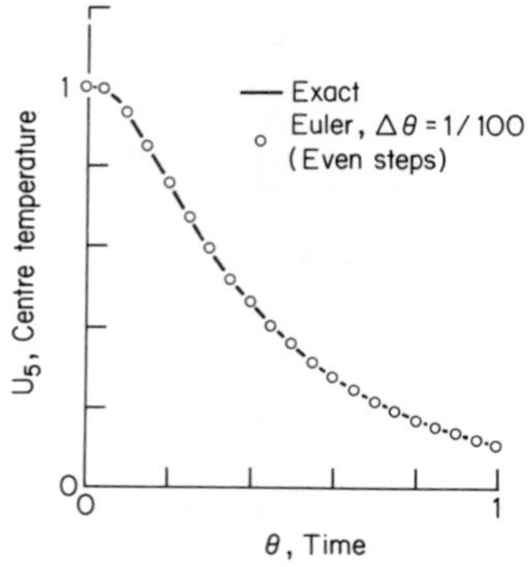

Fig. 9.1 Typical integration assumptions

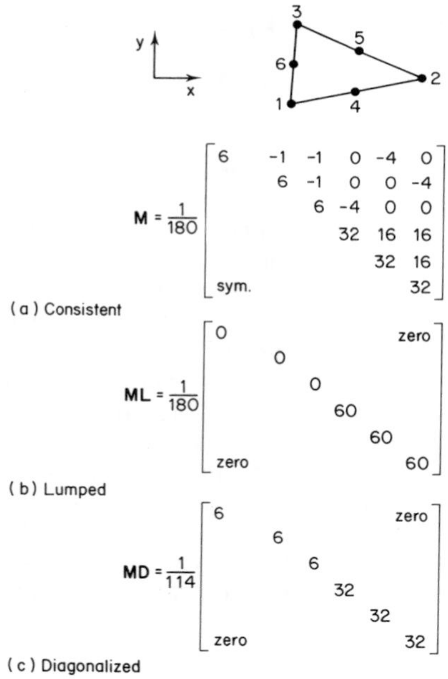

Fig. 9.2 Consistent and diagonal matrices for a quadratic triangle

where q is some constant property per unit volume and **H** denotes the element interpolation functions. This generally can be written as

$$\mathbf{a}^e = Q\mathbf{M}, \tag{9.13}$$

where $Q = qV^e$, and **M** is a symmetric full matrix. Let the sum of the coefficients of the matrix **M** be T, that is

$$T \equiv \sum_i \sum_j M_{ij}. \tag{9.14}$$

In most cases T will be unity but this is not true for axisymmetric integrals. Another quantity of interest is the sum of the diagonal terms of M, i.e.

$$d \equiv \sum_i M_{ii}. \tag{9.15}$$

The most common engineering solution to defining a diagonal matrix is to lump, or sum, all the terms in a given row onto the diagonal of the row and then set the off-diagonal terms to zero. That is, the **lumped matrix L** is defined such that

$$\mathbf{L}_{ii} = \sum_j M_{ij}. \tag{9.16}$$

$$\mathbf{L}_{ij} = 0 \text{ if } i \neq j .$$

Note that doing this does not alter the value of T. Another diagonal matrix, **D**, with the same value of T can be obtained by simply extracting the diagonal of **M** and scaling it by a factor of T/d. That is,

$$D_{ii} = M_{ii}T/d. \tag{9.17}$$

$$D_{ij} = 0 \text{ if } i \neq j .$$

The matrix **D** will be called the diagonalized matrix and **L** the lumped matrix. These matrices are referred to as 'condensed' matrices. For linear simplex elements in two and three dimensions both procedures yield identical diagonal matrices. However, for axisymmetric problems and higher order elements they yield different results and the diagonalized matrix appears to be best in general. This is because for higher order elements the lumped form can introduce zeros or negative numbers on the diagonal. The matrices **M**, **L**, and **D** are illustrated for a quadratic triangle in Fig. 9.2.

9.3 Transient Heat Transfer

For a more detailed example of the above we will apply the Crank-Nicolson method, from Table IX.I, to a simple heat transfer problem. Consider a uniform bar that is initially at a constant temperature, say zero. Let the two ends of the bar be suddenly increased to different temperatures. For simplicity, we will utilize three equal linear elements and a total of four nodes as shown in Fig. 9.3. Since the two end nodes are known we need to determine the time history of the remaining two interior nodes.

The governing partial differential equation which leads to our model Eq. (9.1) is

$$\rho c \frac{\partial R}{\partial t} - K \frac{\partial^2 R}{\partial x^2} = 0 \tag{9.18}$$

where R denotes the temperature, t is time, and the material properties K, ρ, c, are the thermal conductivity, mass density, and the specific heat, respectively. This system also requires initial condition data, i.e. $R(x,t = 0)$, and the boundary conditions as a function of time, say $R(0,t)$ and $R(L,t)$. We have taken the initial condition to be zero, and will arbitrarily set the boundary conditions to be $R_1(t) = 10$ and $R_4(t) = 20$. We have previously shown, in Eq. (4.31), that the element conduction contribution to Eqs. (9.1) and (9.2) is:

$$\mathbf{B}^e = \frac{K^2}{\ell^e} \begin{bmatrix} 1 & -1 \\ -1 & 1 \end{bmatrix} .$$

Myers [16] gives the derivation of the corresponding element heat **capacitance matrix** as

$$\mathbf{A}^e = c^e \rho^e \int_{\ell^e} \mathbf{H}^{e^T} \mathbf{H}^e dx$$

$$= \frac{c^e \rho^e \ell^e}{6} \begin{bmatrix} 2 & 1 \\ 1 & 2 \end{bmatrix} .$$

Assembly of the three equal elements gives Eq. (9.1) as

$$\frac{\rho c \ell}{6}\begin{bmatrix} 2 & 1 & 0 & 0 \\ 1 & 4 & 1 & 0 \\ 0 & 1 & 4 & 1 \\ 0 & 0 & 1 & 2 \end{bmatrix}\dot{\mathbf{R}} + \frac{K}{\ell}\begin{bmatrix} 1 & -1 & 0 & 0 \\ -1 & 2 & -1 & 0 \\ 0 & -1 & 2 & -1 \\ 0 & 0 & -1 & 1 \end{bmatrix}\mathbf{R} = \mathbf{P} \quad .$$

Here the total length of the bar is $L = 3\ell$. The forcing term **P** contains time dependent internal heat sources and external nodal heat sources. Here we assume that these are all zero at the interior nodes.

The above form was used to illustrate the point that Eq. (9.1) requires the storage of two system square matrices. However, we may not want to store them in the above form. When we compare the options in Table IX.I with the actual equations to be solved, i.e. Eq. (9.6), we see that it is probably more efficient to use two alternate square matrix forms. For example, the Crank-Nicolson option needs

$$\mathbf{S} = \mathbf{A}/k + \mathbf{B}/2$$

as well as

$$\mathbf{Q} = \mathbf{A}/k - \mathbf{B}/2$$

to be used in updating the forcing term **F**. Assuming a constant time step, k, and multiplying all terms by $(6k/\rho c \ell)$ lets the Crank-Nicolson method (for $\mathbf{P} = \mathbf{0}$) be written as

$$\mathbf{S}'\mathbf{R}(t + k) = \mathbf{Q}'\mathbf{R}(t) = \mathbf{F}'$$

where

$$\mathbf{S}' = \begin{bmatrix} (2 + b) & (1 - b) & 0 & 0 \\ & (4 + 2b) & (1 - b) & 0 \\ & & (4 + 2b) & (1 - b) \\ \text{sym.} & & & (2 + b) \end{bmatrix}$$

and where $b = 3kK/\rho c \ell^2$. The **Q'** matrix is the same except for a sign change on all the b terms. If we normalize so that $b \equiv 3$ our current numerical model becomes

$$\begin{bmatrix} 5 & -2 & 0 & 0 \\ -2 & 10 & -2 & 0 \\ 0 & -2 & 10 & -2 \\ 0 & 0 & -2 & 5 \end{bmatrix}\mathbf{R}(t + k) = \begin{bmatrix} -1 & 4 & 0 & 0 \\ 4 & -2 & 4 & 0 \\ 0 & 4 & -2 & 4 \\ 0 & 0 & 4 & -1 \end{bmatrix}\mathbf{R}(t) \quad .$$

Of course, these equations must still be modified further since the left side involve R terms that are given boundary conditions in time. When these terms are moved to the right only the second and third rows remain independent:

$$\begin{bmatrix} 10 & -2 \\ -2 & 10 \end{bmatrix} \begin{Bmatrix} R_2 \\ R_3 \end{Bmatrix} = \begin{bmatrix} -2 & 4 \\ 4 & -2 \end{bmatrix} \begin{Bmatrix} R_2(t) \\ R_3(t) \end{Bmatrix}$$

$$- R_1(t + k) \begin{Bmatrix} -2 \\ 0 \end{Bmatrix} - R_4(t + k) \begin{Bmatrix} 0 \\ -2 \end{Bmatrix} .$$

For the results at the first step we substitute the initial conditions on R_2 and R_3, and the new boundary conditions on R_1 and R_4. The resultant right side in that case is

$$F' = \begin{bmatrix} -2 & 4 \\ 4 & -2 \end{bmatrix} \begin{Bmatrix} 0 \\ 0 \end{Bmatrix} - 10 \begin{Bmatrix} -2 \\ 0 \end{Bmatrix} - 20 \begin{Bmatrix} 0 \\ -2 \end{Bmatrix}$$

$$= \begin{Bmatrix} 20 \\ 40 \end{Bmatrix} .$$

Solving for the new values of R_2 and R_3:

$$\begin{Bmatrix} R_2 \\ R_3 \end{Bmatrix} = \frac{1}{96} \begin{bmatrix} 10 & 2 \\ 2 & 10 \end{bmatrix} \begin{Bmatrix} 20 \\ 40 \end{Bmatrix} = \begin{Bmatrix} 2.917 \\ 4.583 \end{Bmatrix} .$$

At the next time step the right side changes to

$$F' = \begin{bmatrix} -2 & 4 \\ 4 & -2 \end{bmatrix} \begin{Bmatrix} 2.917 \\ 4.583 \end{Bmatrix} + \begin{Bmatrix} 20 \\ 0 \end{Bmatrix} + \begin{Bmatrix} 0 \\ 40 \end{Bmatrix} .$$

The corresponding new temperatures are

$$\begin{Bmatrix} R_2 \\ R_3 \end{Bmatrix} = \begin{Bmatrix} 4.271 \\ 5.104 \end{Bmatrix} .$$

Continuing on in this fashion we find that these temperatures eventually reach their steady state values of 13.333 and 16.667, respectively. The unused first and last equations could be used to determine the time dependent end heat flux

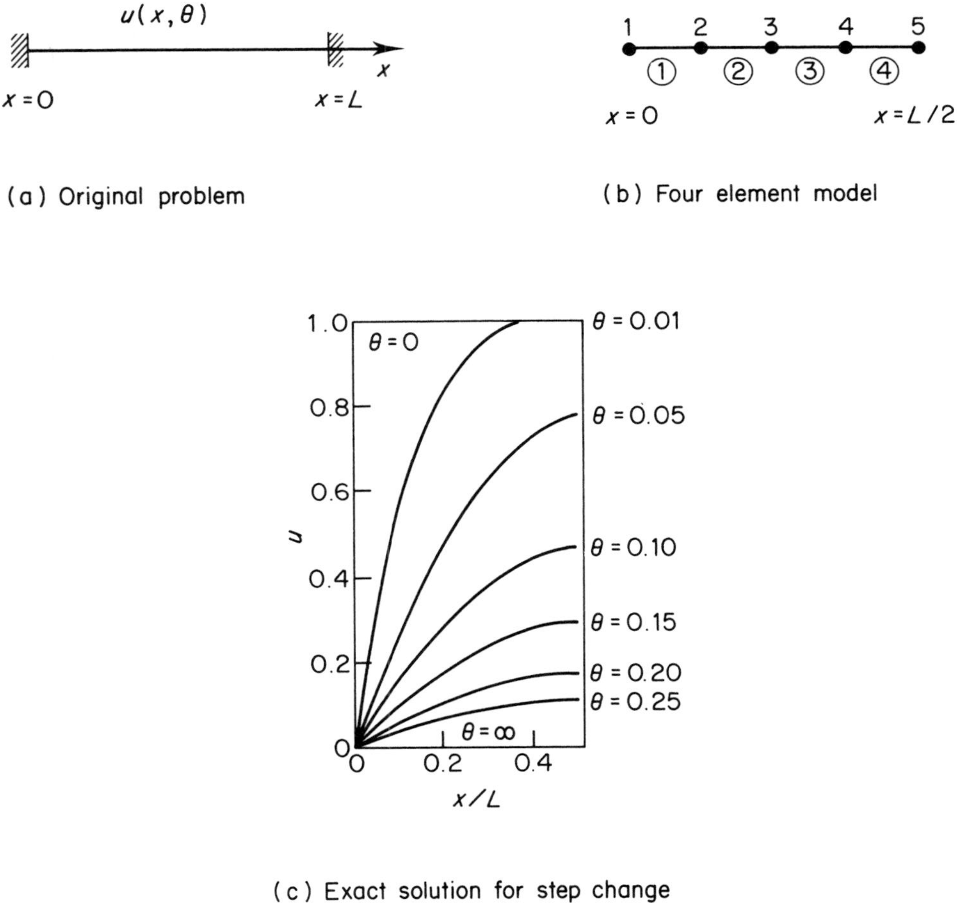

Fig. 9.3 Typical linear transient problem

amounts, P_1 and P_4, necessary to maintain the two end temperatures.

This type of transient calculation is sensitive to the time step size, k. If the time step is too large oscillations can arise in the computed temperatures. If it is too small then a nodal heat flux shock or a sudden temperature increase may lead to physically impossible temperature changes. Both of these effects are caused by incompatible mesh size and time step size. These can be combined with the material properties to form the non-dimensional **Fourier number**:

$$f = \frac{Kk}{\rho c \ell^2} \quad .$$

For transient thermal problems the Fourier number should have a value of approximately unity at all nodes where there is a severe thermal shock. In that case ℓ denotes the smallest distance to another node on the element.

9.4 Exercises

1. Repeat the example in Sec. 9.3 using the diagonal form of the heat capacity matrix.
2. Determine the nodal heat flux reaction at the first node in the above example for:
   a) the first time step, b) the second time step, c) the steady state solution.

10 Error Concepts*

10.1 Introduction

Originally the finite element method was based on physical intuition. The error in the solution was (hopefully) controlled in a similar manner. Today the mathematics of finite elements is precisely defined. Various error analysis proofs are available for numerous classes of analysis problems [6]. Our purpose here is simply to give some insite into the concepts that affect the error in a finite element solution.

10.2 Physical Meanings

10.2.1 One Dimensional Model

To obtain a physical feel for the typical errors involved we consider a one-dimensional model. A hueristic argument will be used. Recall the Taylor's series of a function v at a point x:

$$v(x + h) = v(x) + h\frac{\partial v}{\partial x}(x) + \frac{h^2}{2}\frac{\partial^2 v}{\partial x^2}(x) + \ldots. \qquad (10.1)$$

The objective here is to show that if the third term is neglected then the relations for the linear line element are obtained. That is, the third term is a measure of the interpolation error in the linear element.

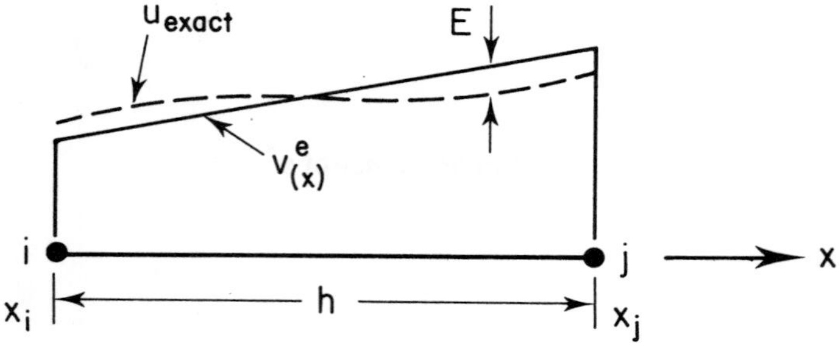

Fig. 10.1 The linear line element

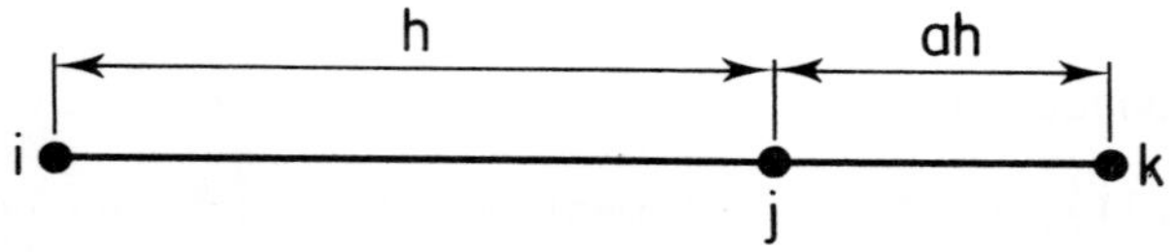

Fig. 10.2 Two adjacent elements

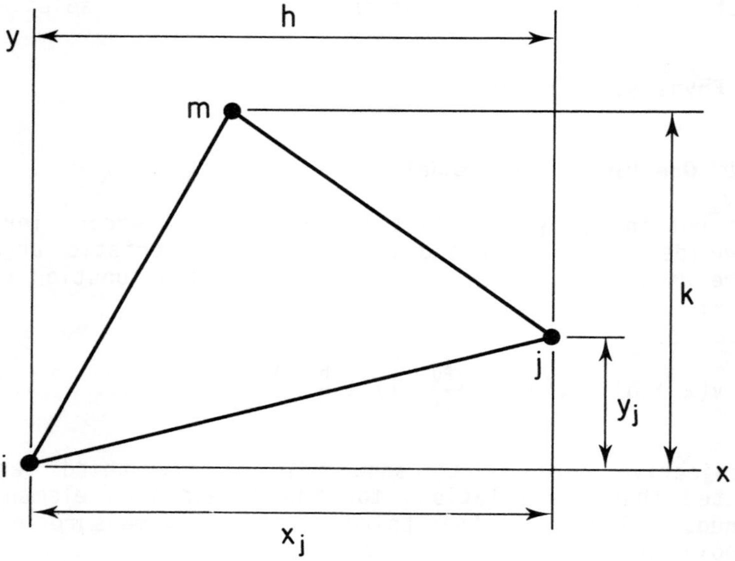

Fig. 10.3 The linear interpolation triangle

For the element shown in Fig. 10.1 use Eq. (10.1) to estimate the function at node j when h is the length of the element:

$$v_j = v_i + h\frac{\partial v}{\partial x}(x_i). \tag{10.2}$$

Solving for the gradient at node i yields

$$\frac{\partial v}{\partial x}(x_i) = (v_j - v_i)/h = \frac{\partial v}{\partial x}(x_j) \tag{10.3}$$

which is the constant previously obtained for the derivative in the linear line element. Thus we can think of this type of element as representing the first two terms of the Taylor series. The omitted third term is a measure of the error associated with the element. Its value is proportional to the product of the second derivative and the square of the element size. That is,

$$E \propto h^2 \frac{\partial^2 v}{\partial x^2} . \tag{10.4}$$

If the exact solution is linear so that the first derivative is constant then the second derivative, $\partial^2 v/\partial x^2$, is zero and there is no error in the element. Otherwise the second derivative and element error do not vanish. If the user wishes to exercise control over this relative error then the element size, h, must be varied.

If we think in terms of the bar element then v and $\partial v/\partial x$ represent the displacement and strain, respectively. The contribution to the error represents the strain gradient (and stress gradient). Therefore we must use our engineering judgment to make the element size, h, small in regions of large strain gradients (stress concentrations). Conversely, where the strain gradients are small we can increase the element size, h, to reduce the computational cost. A similar argument can be stated for the heat conduction problem. Then v is the temperature, $\partial v/\partial x$ describes the temperature gradient (heat flux) and the error is proportional to flux gradient.

If one does not wish to vary the element sizes, h, then to reduce the error one must add higher order polynomial terms of the element interpolation functions so that is present in the element. These two approaches to error control are known as the h-method and the second derivative p-method, respectively.

The previous comments have assumed the use of a uniform mesh. That is, h was the same for all elements in the mesh. Thus the above error discussions have not considered the interaction of adjacent elements. The effects of adjacent element sizes have been evaluated for the case of a continuous bar subject to an axial load. An error term, in the governing differential equation, due to the finite element approximation at node j has been shown to be

$$E = -\frac{h}{3}(1 - a)\frac{\partial^3 v}{\partial x^3}(x_j) + \frac{h^2}{12}\left(\frac{1 + a^3}{1 + a}\right)\frac{\partial^4 v}{\partial x^4}(x_j) + \ldots.,$$

where h is the size of one element and ah is the size of the adjacent element as shown in Fig. (10.2). Here it is seen that for a smooth variation $(a \doteq 1)$ or a uniform mesh $(a = 1)$ then the error in the approximated ODE is of order h squared. However, if the adjacent element sizes differ greatly $(a \neq 1)$ then a larger error of order h is present. This suggests that it is desirable to have a gradual change in element sizes when possible. One should avoid placing a small element adjacent to a large one.

10.2.2 Two-Dimensional Model

In the following chapters two-dimensional problems will be introduced. Here we will briefly outline how the above concepts carry over to two-dimensions. Recall the Taylor expansion of a function, u, at a point (x,y) in two dimensions:

$$u(x + h, y + k) = u(x,y) + \left[h\frac{\partial u}{\partial x}(x,y) + k\frac{\partial u}{\partial y}(x,y)\right] + \frac{1}{2!}\left[h^2\frac{\partial^2 u}{\partial x^2} + 2hk\frac{\partial^2 u}{\partial x\,\partial y} + k^2\frac{\partial^2 u}{\partial y^2}\right] + \ldots. \quad (10.5)$$

The objective here is to show that if the third term is neglected then the relations for a linear interpolation triangle are obtained. That is, we will find that the third term is proportional to the error between the true solution and the interpolated solutions. Consider the triangle shown in Fig. 10.3. Employ Eq. (10.5) to estimate the nodal values u_j and u_m in terms of u_i:

$$u_j = u_i + [x_j \frac{\partial u}{\partial x}(x_i,y_i) + y_m \frac{\partial u}{\partial y}(x_i,y_i)]. \qquad (10.6)$$

The value of $\partial u(x_i,y_i)/\partial x$ can be found by multiplying the first relation by y_m and subtracting the product of y_i and the second relation. The result is

$$\frac{\partial u}{\partial x}(x_i,y_i) = \frac{1}{2A}[u_i(y_j - y_m) + u_m(y_i - y_j) + u_j(y_m - y_i)]$$

where A is the area of the triangle. In a similar manner if we compute this derivative at the other two nodes we obtain

$$\frac{\partial u}{\partial x}(x_j,y_j) = \frac{\partial u}{\partial x}(x_m,y_m) = \frac{\partial u}{\partial x}(x_i,y_i).$$

That is, $\partial u/\partial x$ is a constant in the triangle. Likewise, $\partial u/\partial y$ is a constant.

We will see later (e.g., Eq. (11.13)) that a linear interpolation triangle has constant derivatives that include Eq. (10.7). Thus these common elements will represent the first two terms in Eq. (10.5). Thus the element error is related to the third term:

$$E \;\alpha\; (h^2 \frac{\partial^2 u}{\partial x^2} + 2hk \frac{\partial^2 u}{\partial x \partial y} + k^2 \frac{\partial^2 u}{\partial y^2}) \qquad (10.8)$$

where u is the exact solution and h and k measure the element size in the x and y directions. Once again we would find that these second derivatives are related to the strain and stress gradients. If the strains (e.g., $\varepsilon_x = \partial u/\partial x$) are constant then the error is small or zero.

Before leaving these error comments note that Eq. (10.8) could also be expressed in terms of the ratio of (k/h). This is a measure of the relative shape of the element and it is often called the aspect ratio. For an equilateral element this ratio would be near unity. However, for a long narrow triangle it could be quite large. Generally it is best to keep the aspect ratio near unity (say < 5).

10.3 **Patch Test**

The **patch test** has been proven to be a valid convergence test. It was developed from physical intuition and later written in mathematical forms. The basic concept is fairly simple. Imagine what happens as one introduces a very large, almost infinite, number of elements. Clearly they would become very small in size. If we think of the quantities being integrated to form the element matrices we can make an observation about how the solution would behave in this limit. The integran, such as the strain energy, contains derivative terms that would become constant as the element size shrinks toward zero. Thus, to be valid in the limit the element formulation must be able to yield the correct results in that state. That is, to be assured of convergence one must be able to exactly satisfy the state where the derivatives, in the governing integral statement, take on constant or zero values. This condition can be stated as a mathematical test or as a simple numerical test. The latter option is what we want here. The patch test provides a simple numerical way for a user to test an element, or complete computer program, to verify that it behaves as it should.

We define a patch of elements to be a mesh where at least one node is completely surrounded by elements. Any node of this type is referred to as an interior node. The other nodes are referred to as exterior or perimeter nodes. We will compute the dependent variable at all interior nodes. The derivatives of the dependent variable will be computed in each element. The perimeter nodes are utilized to introduce the boundary conditions required by the test. Assume that the governing integral statement has derivatives of order n. We would like to find boundary conditions that would make those derivatives constant. This can be done by selecting an arbitrary n-th order polynomial function of the global coordinates to describe the dependent variable in the global space that is covered by the patch mesh. Clearly the n-th order derivatives of such a function would be constant as desired. The assumed polynomial is used to define the boundary conditions on the perimeter nodes of the patch mesh.

This is done by substituting the input coordinates at the perimeter nodes into the assumed function and computing the required value of the dependent variable at each such node. Once all of the perimeter boundary conditions are known the solution can be numerically executed. The resulting values of the dependent variable are computed at each interior node. To pass the patch test these computed internal values must agree with the value found when the internal nodal coordinates are substituted into the assumed global polynomial. However, the real test is that when each element is checked the calculated n-th order derivatives must agree with the arbitrarily assumed values used to generate the global function. If an element does not satisfy this test it should not be used. All of the simple elements in this text should pass this test. The patch test can also be used for other purposes. For example, the analyst may wish to distort the element shape and/or change the numerical integration rule to see what effect that has on the numerical accuracy of the patch test.

As a simple elementary example of an analytic solution of the patch test consider the bar element developed in Sec. 4.6. The smallest possible patch is one with two line elements. Such a patch has two exterior nodes and one interior node. For simplicity let the lengths of the two elements be equal and have a value of L.

The governing integral statement, Eq. (4.65), contains only the first derivative of u. Thus an arbitrary linear function can be selected for the patch test since it would have a constant first derivative. Thus select $u(x) = a + bx$ for $0 \leq x \leq 2L$. Assemblying the two element patch gives

$$\frac{AE}{L}\begin{bmatrix} 1 & -1 & 0 \\ -1 & (1+1) & -1 \\ 0 & -1 & 1 \end{bmatrix}\begin{Bmatrix} u_1 \\ u_2 \\ u_3 \end{Bmatrix} = \begin{Bmatrix} F_1 \\ 0 \\ F_3 \end{Bmatrix}$$

where F_1 and F_3 are the unknown reactions associated with the prescribed external displacements. These two exterior patch boundary conditions are obtained by substituting their nodal coordinates in the assumed patch solution

$$u_1 = u(x_1) = a + b(0) = a$$

$$u_3 = u(x_3) = a + b(2L) = a + 2bL.$$

Modifying the assembled equations to include the patch boundary conditions

$$\frac{AE}{L}\begin{bmatrix} 1 & -1 & 0 \\ -1 & 2 & -1 \\ 0 & -1 & 1 \end{bmatrix}\begin{Bmatrix} u_1 = a \\ u_2 \\ u_3 = a + 2bL \end{Bmatrix} = \begin{Bmatrix} F_1 \\ 0 \\ F_3 \end{Bmatrix}$$

$$- \frac{aAE}{L}\begin{Bmatrix} 1 \\ -1 \\ 0 \end{Bmatrix} - \frac{(a + 2bL)AE}{L}\begin{Bmatrix} 0 \\ -1 \\ 1 \end{Bmatrix} .$$

Deleting the first and third rows and columns and retaining the independent second equation gives

$$\frac{2AE}{L} u_2 = 0 + \frac{aAE}{L} + \frac{(a + 2bL)AE}{L} .$$

Thus the internal patch displacement is

$$u_2 = (2a + 2bL)/2 = (a + bL).$$

The value required by the patch test is

$$u(x_2) = (a + bx_2) = (a + bL).$$

This agrees with the computed solution as required to have a valid element. Recall that the element strains are defined as

$$e = 1: \quad \varepsilon = (u_2 - u_1)/L = [(a + bL) - a]/L = b$$

$$e = 2: \quad \varepsilon = (u_3 - u_2)/L = [(a + 2bL) - (a + bL)]/L = b.$$

Thus all element derivatives are constant. However these constants must agree with the constant assumed in the patch. That value is

$$\varepsilon = du/dx = d(a + bx)/dx = b.$$

Thus the patch test is completely satisfied.

A major advantage of the patch test is that it can be carried out numerically. In the above case the constants a and b could have been assigned arbitrary values. Inputting the required numerical values of A, E and L would give a complete numerical description that could be tested in a standard program. Such a procedure also verifies that the computer program satisfies certain minimum requirements. A problem with some elements is that they can pass the patch test for a uniform mesh but fail when an arbitrary irregular mesh is employed. Thus as a general rule one should try to avoid conducting the test with a regular mesh such as the one given in the above example. It would have been wiser to use unequal element lengths such as L and αL where α is an arbitrary constant. The linear element should pass the test for any α.

10.4 Distorted Elements*

From calculus it can be shown that for the change of variables

$$S^e = \int_{\Omega^e} \tilde{B}^e \, da = \int_{\tilde{\Omega}^e} \tilde{B}^e(r,s) \, |J(r,s)| \, drds$$

to be valid it is necessary that the determinant of the Jacobian, $|\mathbf{J}|$, be positive definite. That is, $|\mathbf{J}(r,s)| > 0$. So long as that condition is satisfied we can automatically treat elements that are curved in physical space. This is a very powerful and useful feature of isoparametric elements. However, one can reach the point where the distortion from the shape of the parent (i.e., local coordinate) element adversely affects the solution.

It is possible to establish practical guidelines on the limits of allowable distortion by employing theoretical studies and numerical experiments. We will consider an extreme example of the type of difficulties that can arise due to element distortion. For simplicity we will employ a one-dimensional model. With the two node linear line element there is no extra set of nodes that can be distorted relative to the end nodes. Previously we have seen that this element has a constant Jacobian. Specifically, $|\mathbf{J}| = \ell^e$ which is always positive. Next consider the quadratic line element which has an interior node on the parent element, in Fig. 10.4, that is located at its center ($r = 1/2$). However, the location of the interior node in physical space is determined by the input data, $\mathbf{R}^e$. In other words, the analyst has control of the location of the node in global space.

The interpolation functions for this element, in unit coordinates, were given in Eq. (5.17). Let the global coordinate data be $x_1^e = 0$, $x_2^e = \ell$ and $x_3^e = a\ell$, where 'a' is a constant selected by the analyst. Substituting into the identity

$$x(r) = \mathbf{H}(r)\mathbf{x}^e$$

yields

$$x(r) = 0 + \ell H_2(r) + a\ell H_3(r)$$

or simply

$$x(r) = \ell(4a - 1)r + 2\ell(1 - 2a)r^2. \qquad (10.9)$$

This one-dimensional problem has a Jacobian of

$$J(r) = |\mathbf{J}(r)| = \partial x/\partial r = \ell(4a - 1) + 4\ell(1 - 2a)r. \qquad (10.10)$$

If we locate node 3 so that it has the same relative location as its parent than we set a = 1/2. This places the node at the center of the element. Then from Eq. (10.10)

$$J(r) = \ell(2 - 1) + 0 = \ell$$

so the Jacobian again gives a positive constant for the determinant. There should be no difficulties with this element. Now we distort the element by moving the third node in global space. That is, we assign 'a' any value but 1/2. For example, try a = 1/8. Then $|\mathbf{J}| = (-\ell/2 + 3\ell r)$ which clearly varies with the local coordinate, r. Note, in this example, that $|\mathbf{J}| = 0$ at $r = 1/6$, $|\mathbf{J}| < 0$ for $r < 1/6$ and $|\mathbf{J}| > 0$ only for $r > 1/6$. This is a badly distorted element. A related problem occurs when the global derivatives are computed. They require the inverse of **J** which in turn employs division by $|\mathbf{J}|$. Thus, a distorted element can lead to division by zero. A rule of thumb is to keep the third node within the center fourth of the element. For 2-D and 3-D elements it is also possible to move the node normal to the edge. Similarly, it is possible to distort the original corner angles. Suggested limits are shown in Fig. 10.5.

10.5 **Optimum Derivatives***

In our finite element calculations we often have a need for accurate estimates of the derivatives of the primary variable. For example, in plane stress or plane strain analysis the primary unknowns which we compute are the displacement components of the nodes. However, we often are equally concerned about the strains, and stresses, which are computed from the derivatives of the displacements. Likewise when we model an ideal fluid with a velocity potential we actually have little or no interest in the computed potential; but we are very interested in the velocity components which are the derivatives of the potential.

A logical question at this point is: what location in the element will give me the most accurate estimate of the derivatives? Such points are called optimal points. A heuristic argument for determining their location can be easily presented. Let us begin by recalling some of our previous observations. In Sec. 4.6 we found that our example finite element solution was an **interpolate** solution. That is, it was exact at the node points and approximate elsewhere. Such accuracy is rare but in general one finds that the computed values of the primary variable are most accurate at the node points. Thus for the sake of simplicity we will assume that the element's nodal values are exact.

Recall that we have taken our finite element approximation to be a polynomial of some specific order, say m. If the exact solution is also a polynomial of order m then our finite element solution will be exact everywhere in the element. In addition, the finite element derivative estimates will also be exact. It is rare to have such good luck. In general we must expect our results to only be approximate.

However, we can hope for the next best thing to an exact solution. That would be where the exact solution is a polynomial that is one order higher, say $n = m + 1$, than our finite element polynomial. Let the subscripts E and F denote the exact and finite element solutions, respectively. Consider a one-dimensional formulation in natural coordinates, $-1 \leq a \leq +1$. Then the exact solution could be written as

$$U_E(a) = P_E(a)\,\mathbf{V}_E = [\,1 \quad a \quad a^2 \ldots \quad a^n\,] \begin{Bmatrix} V_1 \\ V_2 \\ \vdots \\ V_n \end{Bmatrix}_E \tag{10.11}$$

and our approximate finite element polynomial solution would be

$$U_F(a) = P_F(a)\,\mathbf{V}_F = [\,1 \quad a \quad a^2 \ldots \quad a^m\,] \begin{Bmatrix} V_1 \\ V_2 \\ \vdots \\ V_m \end{Bmatrix}_F \tag{10.12}$$

where $n = (m + 1)$, as assumed above. In the above $\mathbf{V}_E$ and $\mathbf{V}_F$ represent vectors of unknown constants. In the domain of a typical element these two forms should be almost equal. If we assume that they are equal at the nodes then we can equate $u_E(a_j) = u_F(a_j)$ where a_j is the local coordinate of node j. Then the following identities are obtained.

$$\begin{bmatrix} P_F(a_1) \\ P_F(a_2) \\ \vdots \\ P_F(a_m) \end{bmatrix} \mathbf{V}_F = \begin{bmatrix} P_E(a_1) \\ P_E(a_2) \\ \vdots \\ P_E(a_m) \end{bmatrix} \mathbf{V}_E \tag{10.13}$$

or symbolically

$$\mathbf{A}_F \ \mathbf{V}_F = \mathbf{B}_E \ \mathbf{V}_E \tag{10.14}$$

where B has one more column than the square matrix **A.** Indeed, upon closer inspection we should observe that **B** can be partitioned into a square matrix and column so that

$$\mathbf{B}_E = [\mathbf{A}_F \mid C_E] \tag{10.15}$$

where the column is $\mathbf{C}_E^T = [a_1^n \ a_2^n \ a_3^n \ldots a_m^n]$. If we solve Eq. (10.14) we can relate the finite element constants, V_F, to the exact constants, V_E, at the nodes of the element. Then inverting $\mathbf{A}_F$ Eq. (10.14) gives

$$\mathbf{V}_F = \mathbf{A}_F^{-1}\mathbf{B}_E \ \mathbf{V}_E = [\ \mathbf{I} \mid \mathbf{A}_F^{-1} \ \mathbf{C}_E]\mathbf{V}_E \tag{10.16}$$

or simply

$$\mathbf{V}_F = \mathbf{K}\mathbf{V}_E \tag{10.17}$$

where

$$\mathbf{K} = \mathbf{A}_F^{-1}\mathbf{B}_E$$

is a rectangular matrix. Therefore we can return to Eq. (10.11) and relate everything to $\mathbf{V}_E$. This gives

$$u_F(a) = P_F(a)\mathbf{K}\mathbf{V}_E = P_E(a)\mathbf{V}_E = u_E(a)$$

so for arbitrary V_E one probably has

$$\mathbf{P}_F(a)\mathbf{K} = P_E(a). \tag{10.18}$$

Next the local derivatives are defined equal at an optimal point a_0. That is

$$\frac{\partial \mathbf{P}_F(a_o)}{\partial a}\mathbf{K} = \frac{\partial \mathbf{P}_E(a_o)}{\partial a}. \tag{10.19}$$

Simplifying the algebra this reduces to

$$\frac{\partial \mathbf{P}_F(a_o)}{\partial a}\mathbf{A}_F^{-1} \ \mathbf{C}_E = na_o(n - 1) = (m + 1)a_o m \tag{10.20}$$

which can be solved for the optimal point a_o. Likewise, for the second derivative an optimal location a_s, is found from

$$\frac{\partial^2 \mathbf{P}_F(a_s)}{\partial a^2} \mathbf{A}_F^{-1} \mathbf{C}_E = (m + 1)m\ a_s(m - 1). \qquad (10.21)$$

As an example assume a quadratic finite element in one-dimensional natural coordinates, $-1 \leq a \leq +1$. The exact solution is assumed to be cubic. Therefore

$$\mathbf{P}_F = [1 \quad a \quad a^2\ ]$$
$$\mathbf{P}_E = [1 \quad a \quad a^2 \mid a^3].$$

Selecting the nodes at $a_1 = -1$, $a_2 = 0$, and $a_3 = 1$:

$$\mathbf{A}_F = \begin{bmatrix} 1 & -1 & 1 \\ 1 & 0 & 0 \\ 1 & 1 & 1 \end{bmatrix}, \quad \mathbf{A}_F^{-1} = 1/2 \begin{bmatrix} 0 & 2 & 0 \\ -1 & 0 & 1 \\ 1 & -2 & 1 \end{bmatrix}$$

$$\mathbf{B}_E = \left[\begin{array}{ccc|c} 1 & -1 & 1 & -1 \\ 1 & 0 & 0 & 0 \\ 1 & 1 & 1 & 1 \end{array}\right].$$

Equating the first derivatives at the optimum point a_o.

$$[\ 0 \quad 1 \quad 2a_o \quad 1\ ] = [\ 0 \quad 1 \quad 2a_o \quad 3a_o^2\ ]$$

or simply $1 = 3a_o^2$ so $a_o = \pm 1/\sqrt{3}$. These are the usual Gauss points used in the two point integration rule. Similarly, the **optimal location**, a_s, for the second derivative is found from

$$[\ 0 \quad 0 \quad 2 \quad 0\ ] = [\ 0 \quad 0 \quad 2 \quad 6a_s\ ]$$

so $a_s = 0$, the center of the element. The same sort of procedure can be applied to 2-D and 3-D elements. Generally we find that derivative estimates are least accurate at the nodes. The derivative estimates are usually most accurate at the tabulated integration points. That is indeed fortunate since it means we get a good approximation of the element square matrix. The typical sampling positions for the quadratic elements are shown in Fig. 10.6. It is easy to show that the center of the linear element is the optimum position for sampling the first derivative.

A similar procedure can be utilized for non-isoparametric elements. A common example is the use of the cubic **Hermitian elements.** Recall that this element has two dof at each of the two nodes. Those dof are the function, u, and its first derivative, du/da. The required modification in the procedure is to set

$$u_E(a_j) = u_F(a_j)$$

and

$$\frac{du_E(a_j)}{da} = \frac{du_F(a_j)}{da} \tag{10.22}$$

at each of the two node locations, a_j. Since the element is cubic we assume that the exact solution is of fourth order. Then

$$\mathbf{P}_F = [\ 1 \quad a \quad a^2 \quad a^3\ ]$$

$$\mathbf{P}_E = [\ 1 \quad a \quad a^2 \quad a^3 \quad a^4\ ]$$

and

$$\frac{d\mathbf{P}_F}{da} = [\ 0 \ 1 \quad 2a \quad 3a^2\ ]$$

$$\frac{d\mathbf{P}_E}{da} = [\ 0 \ 1 \quad 2a \quad 3a^2 \quad 4a^3\ ].$$

Setting the nodal parameters equal to the exact solution at the nodes gives

$$\begin{bmatrix} \mathbf{P}_F(a_1) \\ \frac{d\mathbf{P}_F}{da}(a_1) \\ \mathbf{P}_F(a_2) \\ \frac{d\mathbf{P}_F(a_2)}{da} \end{bmatrix} \mathbf{V}_F = \begin{bmatrix} \mathbf{P}_E(a_1) \\ \frac{d\mathbf{P}_E(a_1)}{da} \\ \mathbf{P}_E(a_2) \\ \frac{d\mathbf{P}_E(a_2)}{da} \end{bmatrix} \mathbf{V}_E$$

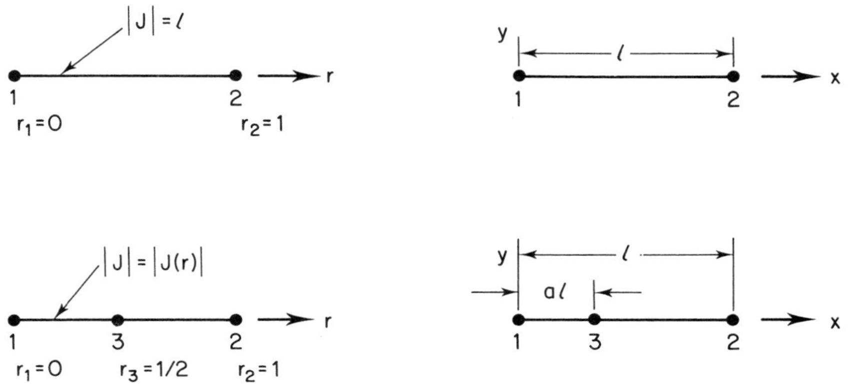

Fig. 10.4 Elements with Constant and Variable Jacobians

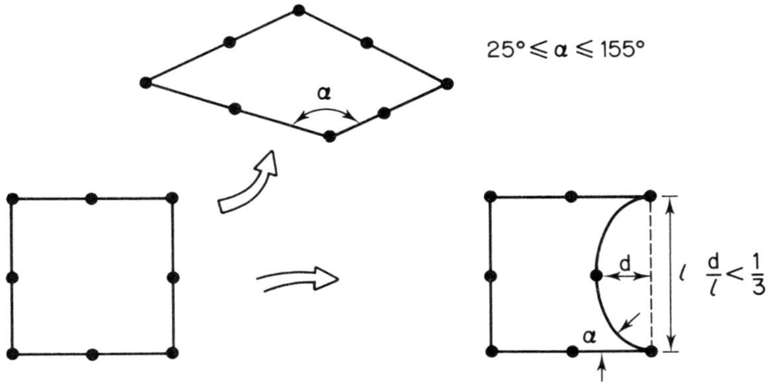

Fig. 10.5 Typical Limits on Distorted Elements

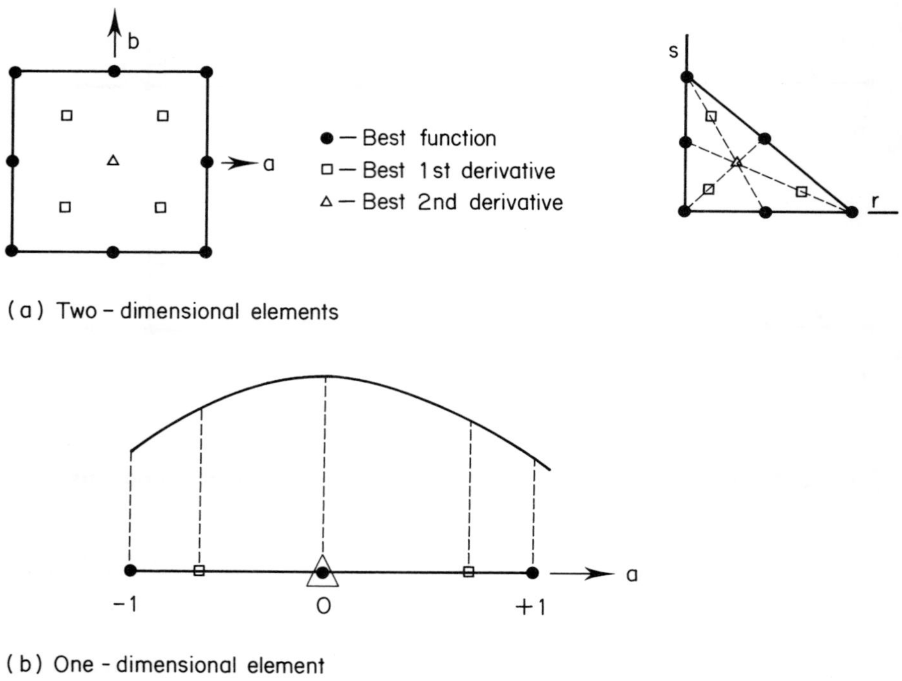

Fig. 10.6 Desired Points for Estimating Derivatives in the Quadratic Elements

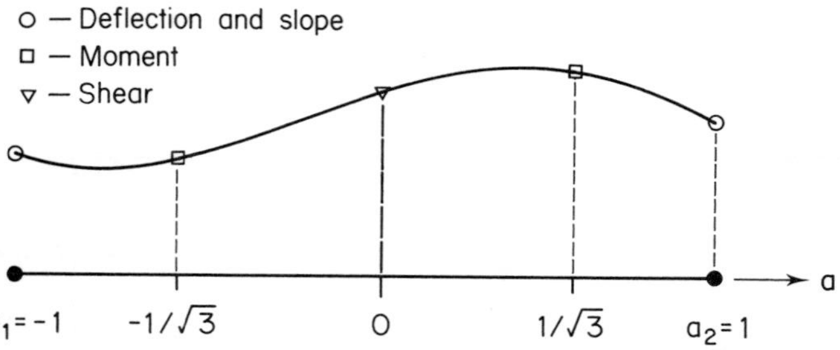

Fig. 10.7 Selective Optimal Output for Hermite Beam Elements

so that for $a_1 = -1$ and $a_2 = +1$:

$$\mathbf{A}_F = \begin{bmatrix} 1 & -1 & 1 & -1 \\ 0 & 1 & -2 & 3 \\ 1 & 1 & 1 & 1 \\ 0 & 1 & 2 & 3 \end{bmatrix}, \mathbf{B}_E = \begin{bmatrix} 1 & -1 & 1 & -1 & 1 \\ 0 & 1 & -2 & 3 & -4 \\ 1 & 1 & 1 & 1 & 1 \\ 0 & 1 & 2 & 3 & 4 \end{bmatrix}.$$

The corresponding generalized form of **K** is

$$K = \left[\begin{array}{cccc|c} 1 & 0 & 0 & 0 & -1 \\ 0 & 1 & 0 & 0 & 0 \\ 0 & 0 & 1 & 0 & 1 \\ 0 & 0 & 0 & 1 & 0 \end{array}\right].$$

Therefore, Eq. (10.18) becomes

$$[\ 1 \quad a \quad a^2 \quad a^3 \quad (-1 + 2a^2)\ ]\ V_E$$
$$= [1 \quad a \quad a^2 \quad a^3 \quad a^4\ ]\ V_E.$$

So the exact and approximate functions are equal only when $(-1 + 2a_o^2) = a_o^4$. That occurs only at the end nodes, $a_o = \pm 1$. Equating the first derivatives at a_f gives $(0 + 4a_f) = 4a_f^3$; which means that they are best at the ends, $a_f = \pm 1$. Let the second derivatives agree at a_s. Then $4 = 12a_s^2$ so that $a_s = \pm 1/\sqrt{3}$, which are the **Gauss points**. Finally, for the third derivatives to agree at a_t requires $0 = 24a_t$, or simply $a_t = 0$; which is the center of the element. This element is often used to model the bending of **beams**. The above analysis suggests that the deflection and slope (u and u') should be output at the two ends. The moment or curvature (u") of an element should be output at the two Gauss points; and the shear (u"') should be output only at the center point of the element. Similar observations carry over to plate bending elements. The conclusions for the beam element are illustrated in Fig. 10.7.

10.6 Exercises

1. Execute, by hand, a patch test of two of the beam elements in Sec. 8.2.2. Let the length of the second element be twice that of the first. Assume that $v(x) = ax^2/2 + bx + c$ where a, b, and c are arbitrary constants.
2. Apply the patch test to the linear heat conduction rod. Assume equal lengths.
3. Utilize a finite element program to apply a patch test for a two dimensional heat conduction problem.

11 INTERPOLATION AND INTEGRATION IN TWO- AND THREE-DIMENSIONS

11.1 Introduction

The previous sections have illustrated the heavy dependence of finite element methods on both spatial interpolation and efficient integrations. In a one-dimensional problem it does not make a great deal of difference if one selects a local or global coordinate system for the interpolation equations. That is because the interelement continuity requirements are relatively easy to satisfy. That is not true in higher dimensions. To obtain practical formulations it is almost essential to utilize local coordinate interpolations. Doing this does require a small amount of additional work in relating the derivatives in the two coordinate systems.

11.2 Unit Coordinate Interpolation

The use of unit coordinates have been previously mentioned in Sec. 5.2. Here some of the procedures for deriving the interpolation functions in unit coordinates will be presented. Consider the three node triangular element shown in Fig. 5.2. The local coordinates of its three nodes are (0,0), (1,0), and (0,1), respectively.

Once again we wish to utilize polynomial functions for our interpolations. In two dimensions the simplest complete polynomial has three constants. Thus this linear function can be related to the three nodal quantities of the element. Assume the polynomial for some quantity, u, is defined as:

$$u^e(r,s) = g_1^{\ e} + g_2^{\ e}r + g_3^{\ e}s = P(r,s)\mathbf{g}^e. \tag{11.1}$$

If it is valid everywhere in the element then it is valid at its nodes. Substituting the local coordinates of a node into Eq. (11.1) gives an identity between the $\mathbf{g}^e$ and a nodal value of u. Establishing these identities at all three nodes gives

$$\begin{Bmatrix} u_1^{\ e} \\ u_2^{\ e} \\ u_3^{\ e} \end{Bmatrix} = \begin{bmatrix} 1 & 0 & 0 \\ 1 & 1 & 0 \\ 1 & 0 & 1 \end{bmatrix} \begin{Bmatrix} g_1^{\ e} \\ g_2^{\ e} \\ g_3^{\ e} \end{Bmatrix}$$

or

$$\mathbf{u}^e = \mathbf{G}\mathbf{g}^e. \tag{11.2}$$

This equation can be solved to yield

$$\mathbf{g}^e = \mathbf{G}^{-1}\mathbf{u}^e \tag{11.3}$$

and

$$u^e(r,s) = P(r,s)\mathbf{G}^{-1}\mathbf{u}^e = \mathbf{H}(r,s)\mathbf{u}^e. \tag{11.4}$$

Here

$$\mathbf{G}^{-1} = \begin{bmatrix} 1 & 0 & 0 \\ -1 & 1 & 0 \\ -1 & 0 & 1 \end{bmatrix} \tag{11.5}$$

and

$$\begin{aligned} H_1(r,s) &= 1 \quad -r \quad -s \\ H_2(r,s) &= \quad\quad r \\ H_3(r,s) &= \quad\quad\quad\quad s\ . \end{aligned} \tag{11.6}$$

Similarly for the unit coordinate quadrilateral in Fig. 5.3 one assumes that

$$u^e(r,s) = g_1^{\ e} + g_2^{\ e}r + g_3^{\ e}s + g_4^{\ e}rs \tag{11.7}$$

so that

$$\mathbf{G} = \begin{bmatrix} 1 & 0 & 0 & 0 \\ 1 & 1 & 0 & 0 \\ 1 & 1 & 1 & 1 \\ 1 & 0 & 1 & 0 \end{bmatrix} \tag{11.8}$$

and

$$H_1(r,s) = 1 - r - s + rs$$
$$H_2 = r - rs$$
$$H_3 = rs$$
$$H_4 = s - rs. \quad (11.9)$$

However, for the quadrilateral it is more common to utilize the natural coordinates shown in Fig. 5.3. In that coordinate system

$$\mathbf{G} = \begin{bmatrix} 1 & -1 & -1 & 1 \\ 1 & 1 & -1 & -1 \\ 1 & 1 & 1 & 1 \\ 1 & -1 & 1 & -1 \end{bmatrix}$$

and the alternate interpolation functions are

$$H_i(a,b) = (1 + aa_i)(1 + bb_i)/4, \quad 1 \leqslant i \leqslant 4 \quad (11.10)$$

where (a_i, b_i) are the local coordinates of node i.

Note that up to this point we have utilized the local element coordinates for interpolation. Doing so makes the geometry matrix, **G**, depend only on element type instead of element number. If we use global coordinates then the geometric matrix, $\mathbf{G}^e$ is always dependent on the element number, e. For example, if Eq. (11.1) is written in physical coordinates then

$$u^e(x,y) = g_1^{\ e} + g_2^{\ e}x + g_3^{\ e}y \quad (11.11)$$

so when the identities are evaluated at each node the result is

$$\mathbf{G}^e = \begin{bmatrix} 1 & x_1^{\ e} & y_1^{\ e} \\ 1 & x_2^{\ e} & y_2^{\ e} \\ 1 & x_3^{\ e} & y_3^{\ e} \end{bmatrix} \quad (11.12)$$

Inverting and simplifying the algebra gives the global coordinate equivalent of Eq. (11.6) for a specific element:

$$H_i^e(x,y) = (a_i^e + b_i^e x + c_i^e y)/2A^e, \quad 1 \leqslant i \leqslant 3 \tag{11.13}$$

where the algebraic constants are

$$\begin{aligned}
a_1^e &= x_2^e y_3^e - x_3^e y_2^e & b_1^e &= y_2^e - y_3^e & c_1^e &= x_3^e - x_2^e \\
a_2^e &= x_3^e y_1^e - x_1^e y_3^e & b_2^e &= y_3^e - y_1^e & c_2^e &= x_1^e - x_3^e \\
a_3^e &= x_1^e y_2^e - x_2^e y_1^e & b_3^e &= y_1^e - y_2^e & c_3^e &= x_2^e - x_1^e
\end{aligned} \tag{11.14}$$

and A^e is the area of the element. That is,

$$A^e = (a_1^e + a_2^e + a_3^e)/2$$

$$A^e = (x_1^e(y_2^e - y_3^e) + x_2^e(y_3^e - y_1^e) + x_3^e(y_1^e - y_2^e))/2. \tag{11.15}$$

These algebraic forms assume that the three local nodes are numbered counter-clockwise from an arbitrarily selected corner. If the topology is defined in a clockwise order then the area, A^e, becomes negative.

It would be natural at this point to attempt to utilize a similar procedure to define the four node quadrilateral in the same manner. For example, if Eq. (11.7) is written as

$$u^e(x,y) = g_1^e + g_2^e x + g_3^e y + g_4^e x\,y \tag{11.16}$$

then the element geometric matrix becomes

$$\mathbf{G}^e = \begin{bmatrix} 1 & x_1^e & y_1^e & x_1^e y_1^e \\ 1 & x_2^e & y_2^e & x_2^e y_2^e \\ 1 & x_3^e & y_3^e & x_3^e y_3^e \\ 1 & x_4^e & y_4^e & x_4^e y_4^e \end{bmatrix}.$$

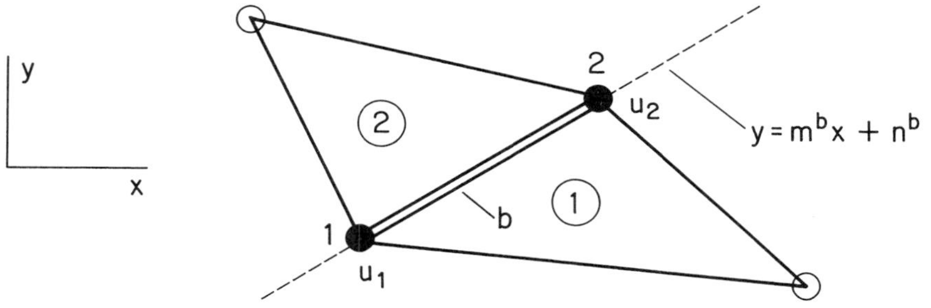

Fig. 11.1 The Common Boundary Between Two Elements

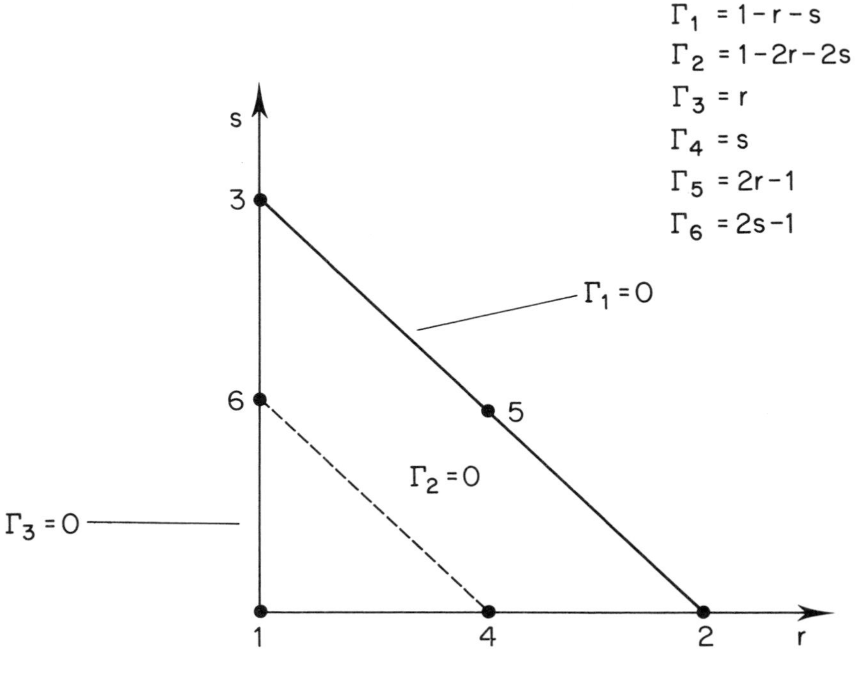

Fig. 11.2 Boundary Curves Through Element Nodes

However, we now find that for a general quadrilateral the inverse of matrix $\mathbf{G}^e$ may not exist. This means that the global coordinate interpolation is in general very sensitive to the orientation of the element in global space.

That is very undesirable. This important disadvantage vanishes only when the element is a rectangle. This global form of interpolation also yields an element that fails to satisfy the required interelement continuity requirements. These difficulties are typical of those that are encountered in two- and three-dimensions when global coordinate interpolation is utilized. Therefore, it is most common to employ the local coordinate mode of interpolation. Doing so also easily allows for the treatment of curvilinear elements. That is done with **isoparametric elements** that will be mentioned later.

At this point it is probably useful to illustrate the lack of continuity that develops in the global coordinate form of the quadrilateral. First consider the three node triangular element and examine the interface or boundary where two elements connect as shown in Fig. 11.1. Along the interface between the two elements one has the geometric restriction that $y = m^b x + n^b$. Recall that the general form of the global coordinate interpolation functions for the triangle is

$$u(x,y) = g_1^e + g_2^e x + g_3^e y$$

where the g_i are element constants. Along the typical interface this reduces to

$$u = g_1^e + g_2^e x + g_3^e (m^b x + n^b)$$

$$u = (g_1^e + n^b g_3^e) + (g_2^e + m^b g_3^e)x$$

or simply

$$u = f_1 + f_2 x.$$

Clearly this shows that the boundary displacement is a linear function of x. The two constants, f_i, could be uniquely determined by noting that $u(x_1) = u_1$ and $u(x_2) = u_2$. Since those two quantities are common to both elements the displacement, $u(x)$, will be continuous between the two elements.

By way of comparison when the same substitution is made in Eq. (11.16) the result for the quadrilateral element is

$$u = g_1^e + g_2^e x + g_3^e (m^b x + n^b) + g_4^e x(m^b x + n^b)$$

$$u = (g_1^e + (g_3^e + g_4^e)n^b) + x(g_2^e + g_3^e m^b)$$

$$+ x^2 g_4^e m^b$$

or simply

$$u = f_1 + f_2 x + f_3 x^2 .$$

This quadratic function can not be uniquely defined by the two constants u_1 and u_2. Therefore, it is not possible to prove that the displacements will be continuous between elements. This is an undesirable feature of quadrilateral elements when formulated in global coordinates.

If the quadrilateral interpolation is given in local coordinates such as Eq. (11.7) or Eq. (11.10) this problem does not occur. On the edge s = 0 Eq. (11.7) reduces to $u = f_1 + f_2 r$. A similar result occurs on the edge s = 1. Likewise for the other two edges $u = f_1 + f_2 s$. Thus in local coordinates the element degenerates to a linear function on any edge and therefore will be uniquely defined by the two shared nodal displacements. In other words, the local coordinate four node quadrilateral will be compatible with elements of the same type and with the three node triangle. The above observations suggest that global coordinates could be utilized for the four node element only so long as it is a rectangle parallel to the global axes.

The extension of the unit coordinates to the three-dimensional tetrahedra illustrated in Fig. 5.2 is straight forward. In the result given below

$$\begin{array}{lcccc}
H_1(r,s,t) = & 1 & -r & -s & -t \\
H_2(r,s,t) = & & r & & \\
H_3(r,s,t) = & & & s & \\
H_4(r,s,t) = & & & & t \ .
\end{array}$$

the dashed partitions illustrate that the 2-D and 1-D forms are contained in the three-dimensional form. This concept was suggested by the topology relations shown in Fig. 5.2.

The unit coordinate interpolation is easily extended to quadratic, cubic, or higher interpolation. The procedure employed to generate Eq. (11.6) can be employed. An alternate geometric approach can be utilized. We want to generate an interpolation function, H_i, that vanishes at the j-th node when $i \neq j$. Such a function can be obtained by taking the products of the equations of selected curves through the nodes on the element. For example, let

$$H_1(r,s) = C_1 \Gamma_1 \Gamma_2$$

where the Γ_i are the equations of the lines are shown in Fig. 11.2 and where C_1 is a constant chosen so that $H_1(r_1,s_1) = 1$. This yields

$$H_1 = (1 - 3r - 3s + 2r^2 + 4rs + 2s^2).$$

Similarly, letting $H_4 = C_4 \Gamma_1 \Gamma_3$ gives $C_4 = 4$ and $H_4 = 4r(1 - r - s)$. This type of procedure is usually quite straightforward. However, There are times when there is not a unique choice of products, and then care must be employed to select the proper products. The resulting two-dimensional interpolation functions for the quadratic triangle are

$$\begin{aligned}
H_1(r,s) &= 1 \quad -3r \quad +2r^2 \quad -3s \quad +4rs \quad +2s^2 \\
H_2(r,s) &= \quad -r \quad +2r^2 \\
H_3(r,s) &= \quad -s \quad +2s^2 \\
H_4(r,s) &= \quad 4r \quad -4r^2 \quad -4rs \\
H_5(r,s) &= \quad 4rs \\
H_6(r,s) &= \quad 4s \quad -4rs \quad -4s^2. \qquad (11.17)
\end{aligned}$$

Once again, it is possible to obtain the one-dimensional quadratic interpolation on a typical edge by setting $s = 0$. Figure 11.3 shows the shape of the typical interpolation functions for a triangular element.

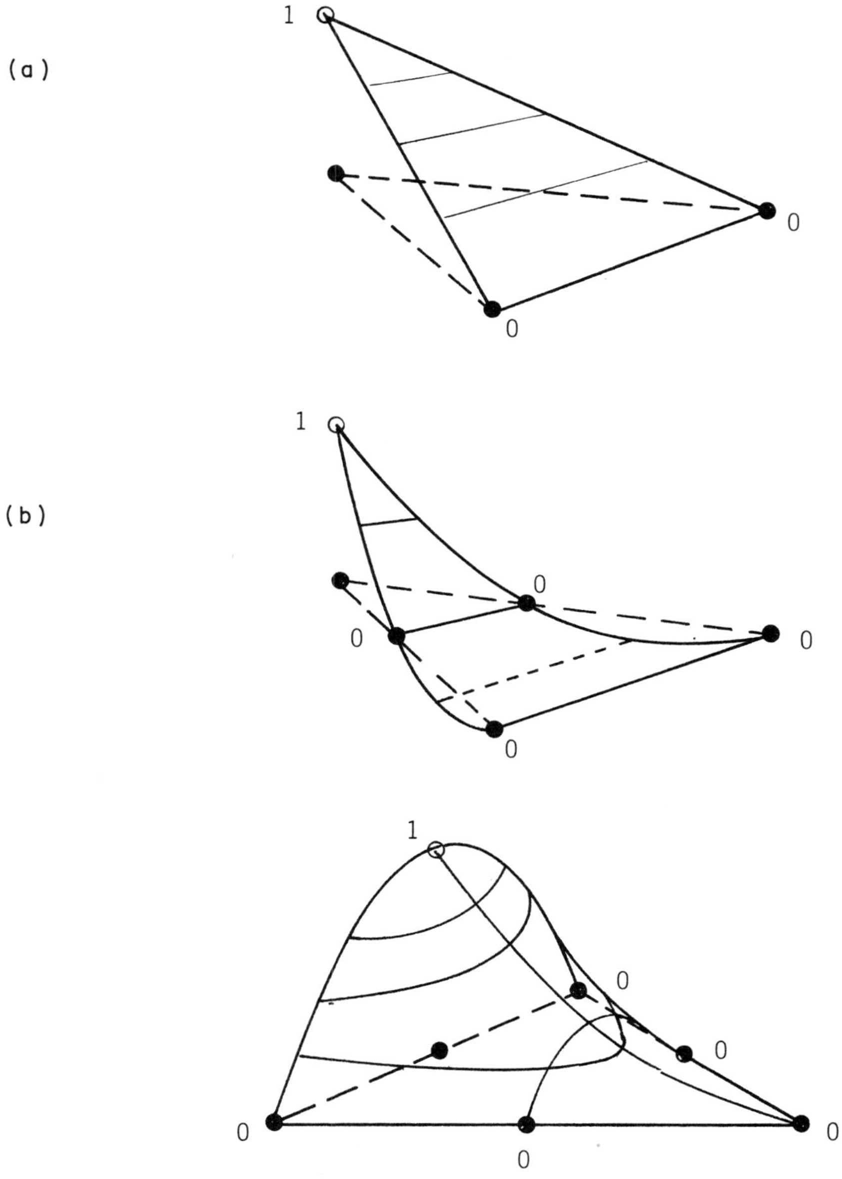

Fig. 11.3 Typical interpolation functions for a) linear and b) quadratic triangles

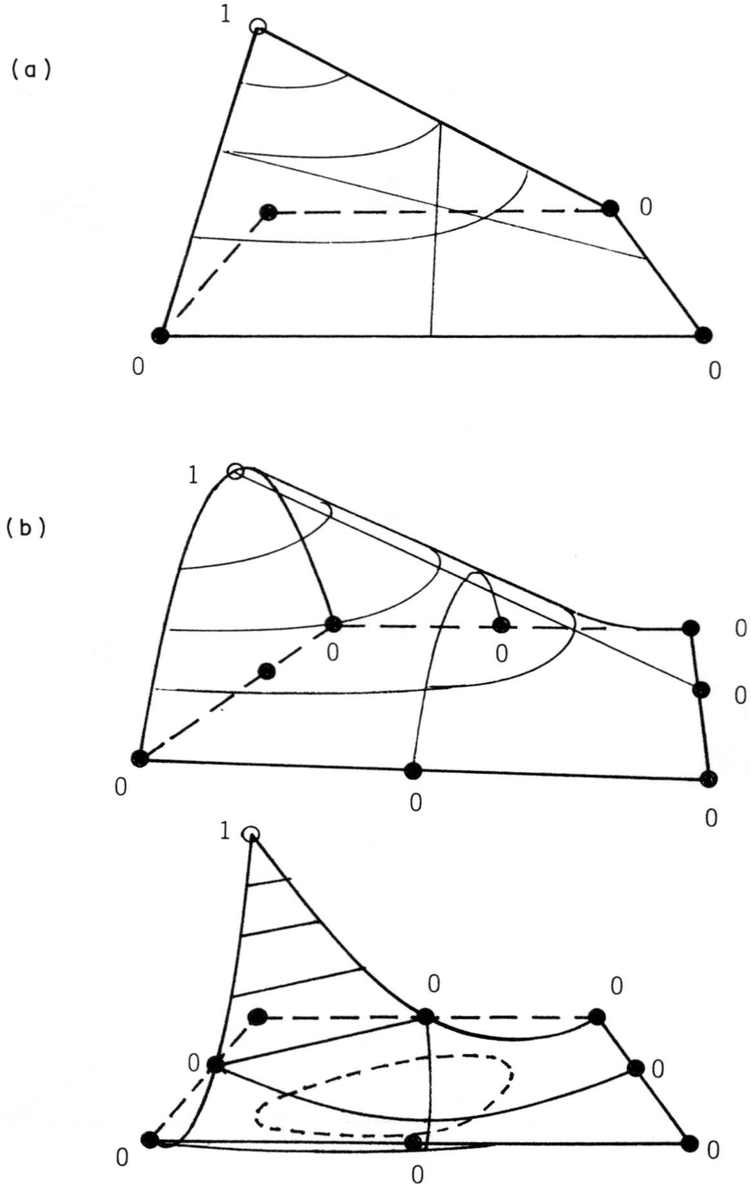

Fig. 11.4 Typical interpolation functions are the a) four and b) eight node quadrilaterals

11.3 Natural Coordinates

The natural coordinate formulations for the interpolation functions can be generated in a similar manner as was illustrated in Eq. (11.10). However, the inverse geometric matrix, G^{-1}, may not be unique. In such cases the commonly used functions are obtained when the inverse is obtained in a least squares sense. However, the most common functions have been known for several years and will be presented here in two groups. They are generally denoted as Lagrangian elements and as the Serendipity elements. For the four node quadrilateral element both forms yield Eq. (11.10). This is known as the bi-linear quadrilateral since it has linear interpolation on its edges and a bi-linear (incomplete quadratic) interpolation on its interior. This element is easily extended to the tri-linear hexahedra of Fig. 5.3. Its resulting interpolation functions are

$$H_i(a,b,c) = (1 + aa_i)(1 + bb_i)(1 + cc_i)/8, \qquad (11.18)$$

for $1 \leqslant i \leqslant 8$ where (a_i, b_i, c_i) are the local coordinates of node i. On a given face, eq. $c = \pm 1$, these degenerate to the four functions in Eq. (11.10) and four zero terms.

For quadratic (or higher) edge interpolation, the Lagrangian and Serendipity elements are different. The Serendipity interpolation functions for the corner quadratic nodes are

$$H_i(a,b) = (1 + aa_i)(1 + bb_i)(aa_i + bb_i - 1)/4, \qquad (11.19)$$

where $1 \leqslant i \leqslant 4$ and for the midside nodes

$$H_i(a,b) = a_i^2(1 - b^2)(1 + a_i a)/2 + b_i^2(1 - a^2)(1 + b_i b)/2, \quad 5 \leqslant i \leqslant 8 . \qquad (11.20)$$

Other members of this family are shown in Fig. 11.4. The Lagrangian functions are obtained from the products of the one-dimensional equations. The resulting quadratic functions are

$$H_1(a,b) = (a^2 - a)(b^2 - b)/4 \qquad H_6(a,b) = (a^2 + a)(1 - b^2)/2$$
$$H_2(a,b) = (a^2 + a)(b^2 - b)/4 \qquad H_7(a,b) = (1 - a^2)(b^2 + b)/2$$
$$H_3(a,b) = (a^2 + a)(b^2 + b)/4 \qquad H_8(a,b) = (a^2 - a)(1 - b^2)/2$$
$$H_4(a,b) = (a^2 - a)(b^2 + b)/4 \qquad H_9(a,b) = (1 - a^2)(1 - b^2)$$
$$H_5(a,b) = (1 - a^2)(b^2 - b)/2 \qquad .$$

The typical shapes of these functions are shown in Fig. 11.5. The equations for the common serendipity elements in 2-D are given in Fig. 11.6 and their extensions to the 3-D family are given in Fig. 11.7.

It is possible to mix the order of interpolation on the edges of an element. Fig. 11.8 illustrates the Serendipity interpolation functions [1] for quadrilateral element that can be either linear, quadratic, or cubic on any of its four sides. Such an element is often referred to as a **transition element.** The typical polynomial terms contributing to the most common two-dimensional elements are illustrated in Figs. 11.9 and 11.10. These figures also illustrate the usual number of nodes associated with these elements. Of course, the elements with the single constant terms are not used in practice.

11.4 Isoparametric Elements

By introducing local coordinates to formulate the element interpolation functions we were able to satisfy certain continuity requirements that could not be satisfied by global coordinate interpolation. We will soon see that a useful by-product of this approach is the ability to treat elements with curved edges. At this point there may be some concern about how one relates the local coordinates to the global coordinates. This must be done since the governing integral is presented in global (physical) coordinates and it involves derivatives with respect to the global coordinates. This can be accomplished with the popular **isoparametric elements.**

11.4.1 Basic Concepts

Isoparametric elements utilize a local coordinate system to formulate the element matrices. The local coordinates, say

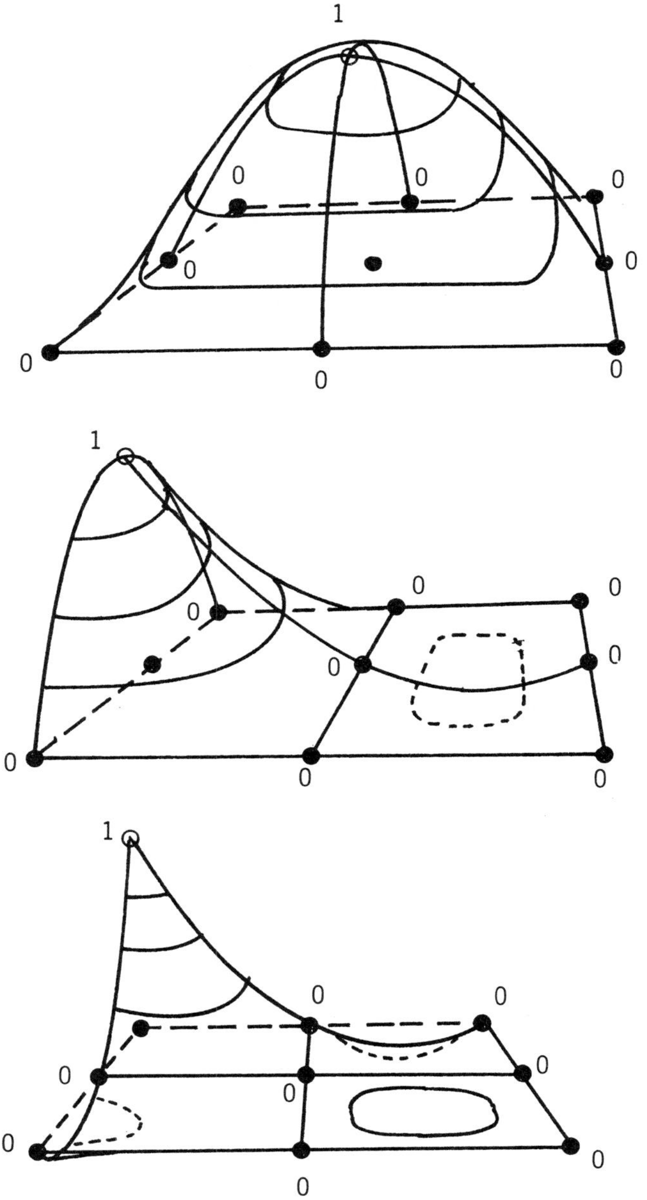

Fig. 11.5 Typical interpolation functions for the nine node quadrilateral

Node Location		Interpolation Functions	Mesh Figure
a_i	b_i	$H_i(a, b)$	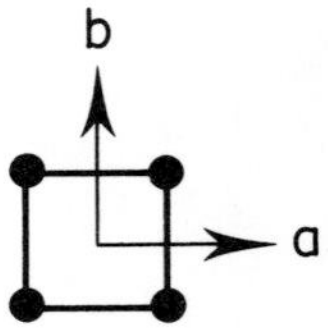
±1	±1	$(1 + aa_i)(1 + bb_i)/4$	
±1	±1	$(1 + aa_i)(1 + bb_i)(aa_i + bb_i - 1)/4$	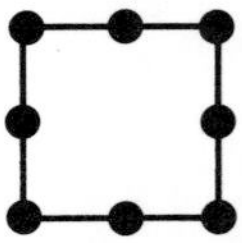
±1	0	$(1 + aa_i)(1 - b^2)/2$	
0	±1	$(1 + bb_i)(1 - a^2)/2$	
±1	±1	$(1 + aa_i)(1 + bb_i)[9(a^2 + b^2) - 10]/32$	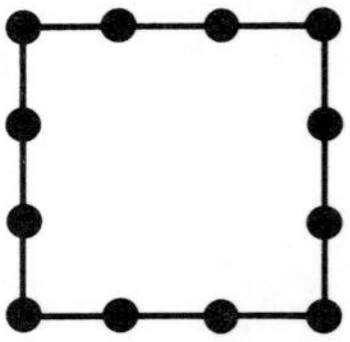
±1	±1/3	$9(1 + aa_i)(1 - b^2)(1 + 9bb_i)/32$	
±1/3	±1	$9(1 + bb_i)(1 - a^2)(1 + 9aa_i)/32$	
±1	±1	$(1 + aa_i)(1 + bb_i)[4(a^2 - 1)aa_i + 4(b^2 - 1)bb_i + 3aba_ib_i]/12$	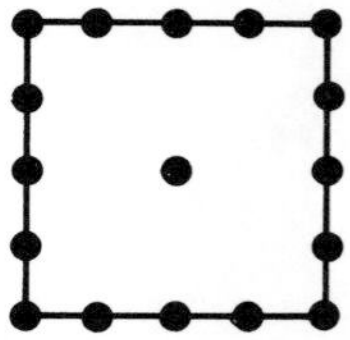
±1	0	$2(1 + aa_i)(b^2 - 1)(b^2 - aa_i/4)$	
0	±1	$2(1 + bb_i)(a^2 - 1)(a^2 - bb_i/4)$	
±1	±1/2	$4(1 + aa_i)(1 - b^2)(b^2 + bb_i)/3$	
±1/2	±1	$4(1 + bb_i)(1 - a^2)(a^2 + aa_i)/3$	
0	0	$(a^2 - 1)(b^2 - 1)$	

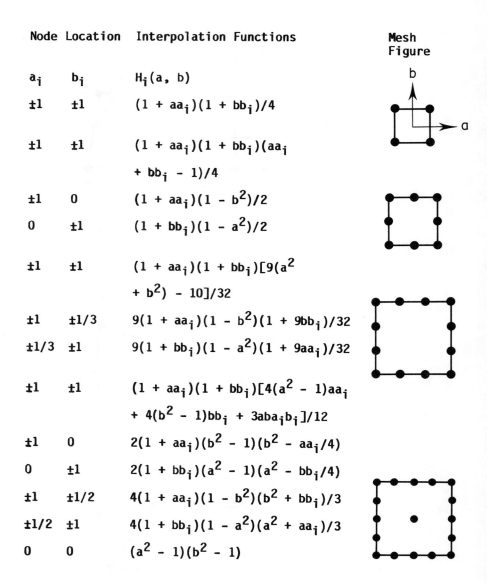

Fig. 11.6 Serendipity Quadrilaterals in Natural Coordinates

Node Location			Interpolation Functions	Mesh Figure
a_i	b_i	c_i	$H_i(a, b, c)$	
±1	±1	±1	$(1 + aa_i)(1 + bb_i)(1 + cc_i)/8$	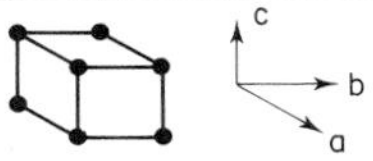
±1	±1	±1	$(1 + aa_i)(1 + bb_i)(1 + cc_i)(aa_i + bb_i + cc_i - 2)/8$	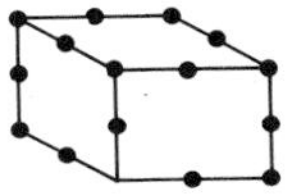
0	±1	±1	$(1 - a^2)(1 + bb_i)(1 + cc_i)/4$	
±1	0	±1	$(1 - b^2)(1 + aa_i)(1 + cc_i)/4$	
±1	±1	0	$(1 - c^2)(1 + aa_i)(1 + bb_i)/4$	
±1	±1	±1	$(1 + aa_i)(1 + bb_i)(1 + cc_i)[9(a^2 + b^2 + c^2) - 19]/64$	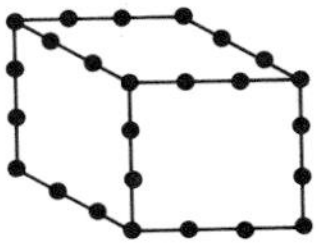
±1/3	±1	±1	$9(1 - a^2)(1 + 9aa_i)(1 + bb_i)(1 + cc_i)/64$	
±1	±1/3	±1	$9(1 - b^2)(1 + 9bb_i)(1 + aa_i)(1 + cc_i)/64$	
±1	±1	±1/3	$9(1 - c^2)(1 + 9cc_i)(1 + bb_i)(1 + aa_i)/64$	

Fig. 11.7 Serendipity Hexahedra in Natural Coordinates

r, s, and t, are usually dimensionless and range from 0 to 1, or from -1 to 1. The latter range is usually preferred since it is directly compatible with the definition of abscissa utilized in numerical integration by Gaussian quadratures. The elements are called isoparametric since the same (iso) local coordinate parametric equations (interpolation functions) used to define any quantity of interest within the elements are also utilized to define the global coordinates of any point within the element in terms of the global spatial coordinates of the nodal points. Let the global spatial coordinates again be denoted by x, y, and z. Let the number of nodes per element be N. For simplicity, consider a single scalar quantity of interest, say V(r,s,t). The value of this variable at any local point (r,s,t) within the element is assumed to be defined by the values at the N nodal points of the element (V_i^e, $1 \leq i \leq N$), and a set of interpolation functions ($H_i(r,s,t)$, $1 \leq i \leq N$). That is,

$$V(r,s,t) = \sum_{i=1}^{N} H_i(r,s,t) V_i^e,$$

or symbolically,

$$V(r,s,t) = \mathbf{H}\mathbf{V}^e, \tag{11.22}$$

where **H** is a row vector. Generalizing this concept, the global coordinates are defined as

$$x(r,s,t) = \mathbf{H}\mathbf{x}^e, \; y = \mathbf{H}\mathbf{y}^e, \; z = \mathbf{H}\mathbf{z}^e.$$

Programming considerations make it desirable to write the last three relations as a position matrix, **R**, written in a partitioned form

$$\mathbf{R}(r,s,t) = \mathbf{H}(r,s,t)\,\mathbf{R}^e = \mathbf{H}[\mathbf{x}^e \mathbf{y}^e \mathbf{z}^e] \tag{11.23}$$

where the last matrix simply contains the spatial coordinates of the N nodal points incident with the element. To illustrate a typical two-dimensional isoparametric element, consider a quadrilateral element with nodes at the four corners, as shown in Fig. 5.3. The global coordinates and local coordinates of a typical corner, i, are (x_i, y_i), and (r_i, s_i) respectively. The following local coordinate interpolation functions (shape functions) have been developed earlier for this element:

$$H_i(r,s) = \frac{1}{4}(1 + rr_i)(1 + ss_i), \qquad 1 \leq i \leq 4.$$

Recall that

$$V(r,s) = H(r,s)\mathbf{V}^e$$

$$= [H_1 \quad H_2 \quad H_3 \quad H_4] \begin{Bmatrix} V_1 \\ V_2 \\ V_3 \\ V_4 \end{Bmatrix}^e$$

and

$$x(r,s) = H(r,s)\mathbf{x}^e,$$

$$y(r,s) = H(r,s)\mathbf{y}^e.$$

Note that along an edge of the element ($r = \pm 1$, or $s = \pm 1$) these interpolation functions become linear and thus any of these three quantities can be uniquely defined by the two corresponding nodal values on that edge. If the adjacent element is of the same type (linear on the boundary), then these quantities will be continuous between elements since their values are uniquely defined by the shared nodal values on that edge. Since the variable of interest, V, varies linearly on the edge of the element, it is called the linear isoparametric quadrilateral although the interpolation functions are bilinear inside the element.

For future reference, note that if one can define the interpolation functions in terms of the local coordinates then one can also define their partial derivatives with respect to the local coordinate system. For example, the local derivatives of the shape functions of the above element are

$$\frac{\partial H_i(r,s)}{\partial r} = \frac{1}{4} r_i(1 + ss_i),$$

$$\frac{\partial H_i(r,s)}{\partial s} = \frac{1}{4} s_i(1 + rr_i).$$

In the three dimensions, let the array containing the local derivatives of the interpolation functions be denoted by $\mathbf{\Delta}$, a 3 by N matrix, where

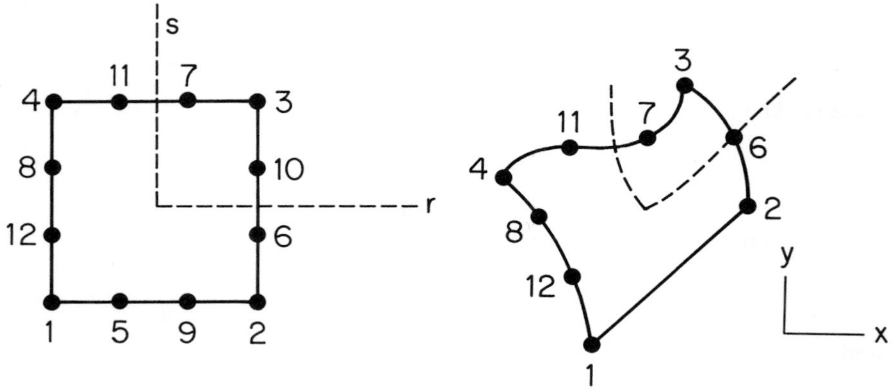

Fig. 11.8 Linear to Cubic Transition Quadrilateral

If Cubic Side: i = 5, 9, or 6, 10 or 7, 11 or 8, 12

$$H_i(r,s) = (1 - s^2)(1 + 9ss_i)(1 + rr_i)9/32$$

$$H_i(r,s) = (1 - r^2)(1 + 9rr_i)(1 + ss_i)9/32$$

If Quadratic Side: i = 5, 6, 7, or 8

$$H_i(r,s) = (1 + rr_i)(1 - s^2)/2$$

$$H_i(r,s) = (1 + ss_i)(1 - r^2)/2$$

$$H_j = 0, \; j = i + 4$$

If Linear Side:

$$H_j = H_k = 0, \; j = i + 4, \; k = i + 8, \; i = 1, 2, 3, \text{ or } 4$$

At Corners: i = 1, 2, 3, 4

$$H_i(r,s) = (P_r + P_s)(1 + ss_i)/4$$

Order of Side	P_r, $s_i = \pm 1$	P_s, $r_i = \pm 1$
Linear	1/2	1/2
Quadratic	$rr_i - 1/2$	$ss_i - 1/2$
Cubic	$(9r^2 - 5)/8$	$(9s^2 - 5)/8$

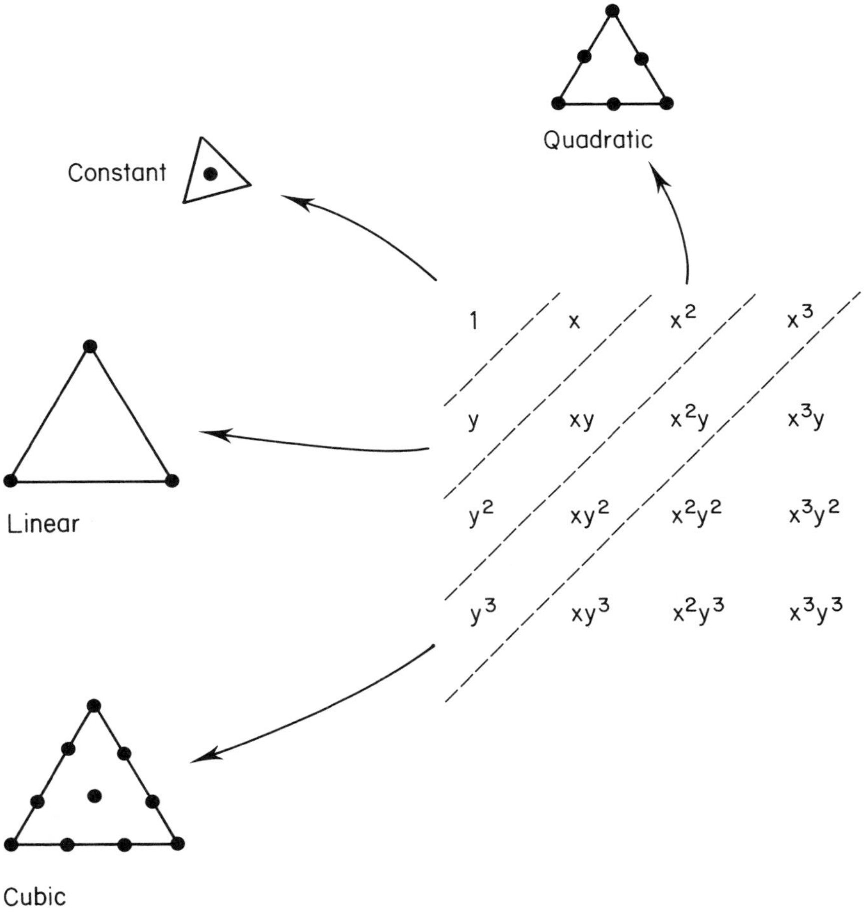

Fig. 11.9 Reasonable Approximations for Manual Input for Quadratic and Cubic Elements

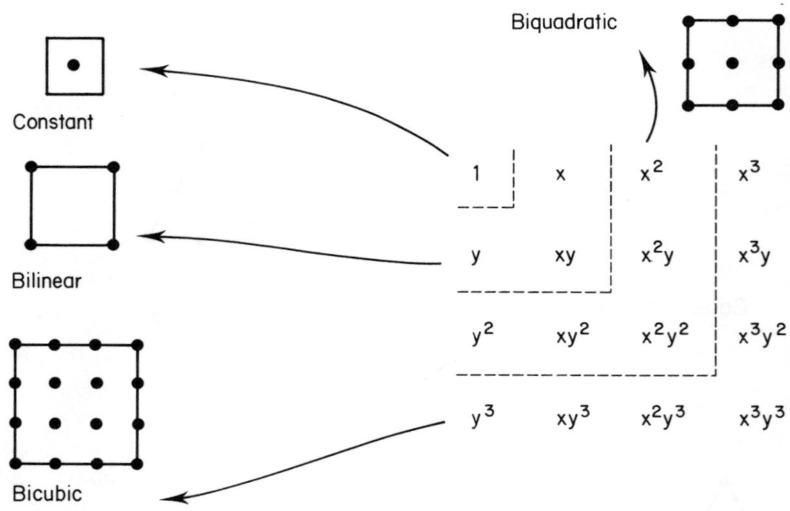

(a) Lagrangian forms

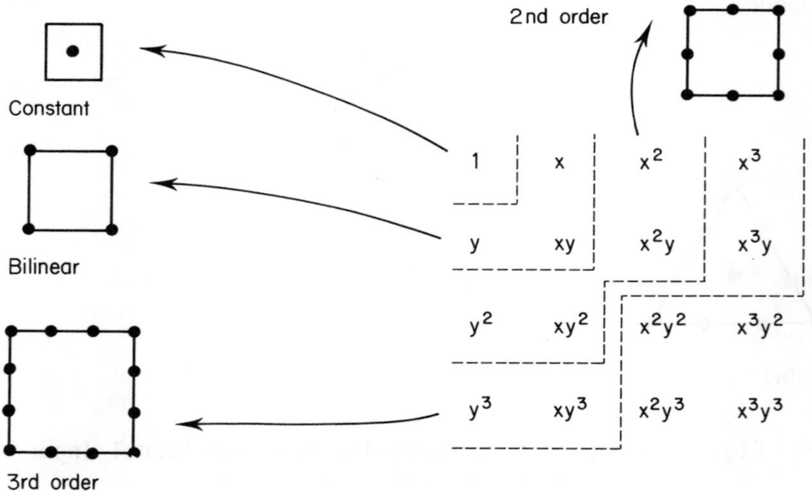

(b) Serendipity forms

Fig. 11.10 The Common Quadrilaterals

$$\Delta(r,s,t) = \begin{bmatrix} \frac{\partial}{\partial r} \mathbf{H} \\ \frac{\partial}{\partial s} \mathbf{H} \\ \frac{\partial}{\partial t} \mathbf{H} \end{bmatrix} = \boldsymbol{\partial}_{\ell}\mathbf{H}$$

Although x, y, and z can be defined in an isoparametric element in terms of the local coordinates, r, s, and t, a unique inverse transformation is not needed. Thus one usually does not define r, s, and t in terms of x, y, and z. What one must have, however, are the relations between derivatives in the two coordinate systems. From calculus, it is known that the derivatives are related by the **Jacobian.** Recall that from the chain rule of calculus one can write, in general,

$$\frac{\partial}{\partial r} = \frac{\partial}{\partial x}\frac{\partial x}{\partial r} + \frac{\partial}{\partial y}\frac{\partial y}{\partial r} + \frac{\partial}{\partial z}\frac{\partial z}{\partial r}$$

with similar expressions for $\partial/\partial s$ and $\partial/\partial t$. In matrix form these become

$$\left\{ \begin{matrix} \frac{\partial}{\partial r} \\ \frac{\partial}{\partial s} \\ \frac{\partial}{\partial t} \end{matrix} \right\} = \begin{bmatrix} \frac{\partial x}{\partial r} & \frac{\partial y}{\partial r} & \frac{\partial z}{\partial r} \\ \frac{\partial x}{\partial s} & \frac{\partial y}{\partial s} & \frac{\partial z}{\partial s} \\ \frac{\partial x}{\partial t} & \frac{\partial y}{\partial t} & \frac{\partial z}{\partial t} \end{bmatrix} \left\{ \begin{matrix} \frac{\partial}{\partial x} \\ \frac{\partial}{\partial y} \\ \frac{\partial}{\partial z} \end{matrix} \right\} \qquad (11.24)$$

where the square matrix is called the Jacobian. Symbolically, one can write the derivatives of a quantity, such as V(r,s,t), which for convenience is written as V(x,y,z) in the global coordinate system, in the following manner

$$\boldsymbol{\partial}_{\ell}V = \mathbf{J}(r,s,t)\,\boldsymbol{\partial}_{g}V,$$

where $\mathbf{J}$ is the Jacobian matrix and where the subscripts ℓ, and g have been introduced to denote local and global derivatives, respectively. Similarly, the inverse relation is

$$\boldsymbol{\partial}_{g}V = \mathbf{J}^{-1}\boldsymbol{\partial}_{1}V. \qquad (11.25)$$

Thus, to evaluate global and local derivatives, one must be able to establish $\mathbf{J}$ and $\mathbf{J}^{-1}$. In practical application, these two quantities usually are evaluated numerically. Consider the first term in $\mathbf{J}$:

$$\frac{\partial x}{\partial r} = \frac{\partial}{\partial r} \mathbf{H}\mathbf{x}^e .$$

Similarly for any component in Eq. (11.23)

$$\frac{\partial \mathbf{R}}{\partial r} = \frac{\partial}{\partial r} \mathbf{H}\mathbf{R}^e .$$

Repeating for all local direction we find the identity that

$$\begin{bmatrix} \frac{\partial x}{\partial r} & \frac{\partial y}{\partial r} & \frac{\partial z}{\partial r} \\ \frac{\partial x}{\partial s} & \frac{\partial y}{\partial s} & \frac{\partial z}{\partial s} \\ \frac{\partial x}{\partial t} & \frac{\partial y}{\partial t} & \frac{\partial z}{\partial t} \end{bmatrix} = \begin{bmatrix} \frac{\partial}{\partial r} & \mathbf{H} \\ \frac{\partial}{\partial s} & \mathbf{H} \\ \frac{\partial}{\partial t} & \mathbf{H} \end{bmatrix} \mathbf{R}^e$$

or, in symbolic form,

$$\mathbf{J}^e(r,s,t) = \mathbf{\Delta}(r,s,t)\mathbf{R}^e . \qquad (11.26)$$

This numerically defines the Jacobian matrix, $\mathbf{J}$, at a local point inside a typical element in terms of the spatial coordinates of the element's nodes, $\mathbf{R}^e$, which is referenced by the name **COORD** in the subroutines, and the local derivatives, $\mathbf{\Delta}$, of the interpolation functions, $\mathbf{H}$. Thus, at any point (r,s,t) of interest, such as a numerical integration point, it is possible to define the values of $\mathbf{J}$, $\mathbf{J}^{-1}$, and the determinant of the Jacobian, $|\mathbf{J}|$.

Of course, one can also establish the algebraic form of $\mathbf{J}$. For simplicity consider the three node triangle in two dimensions. From Eq. (11.6) we note that the local derivatives of $\mathbf{H}$ are

$$\mathbf{\Delta} = \begin{bmatrix} \partial \mathbf{H}/\partial r \\ \partial \mathbf{H}/\partial s \end{bmatrix} = \begin{bmatrix} -1 & 1 & 0 \\ -1 & 0 & 1 \end{bmatrix} . \qquad (11.27)$$

Thus the element has constant local derivatives since no functions of the local coordinates remain. The Jacobian matrix is

$$\mathbf{J} = \begin{bmatrix} \partial x/\partial r & \partial y/\partial r \\ \partial x/\partial s & \partial y/\partial s \end{bmatrix} .$$

Employing Eq. (11.26) for a specific element:

$$\mathbf{J}^e = \Delta\mathbf{R}^e = \begin{bmatrix} -1 & 1 & 0 \\ -1 & 0 & 1 \end{bmatrix} \begin{bmatrix} x_1 & y_1 \\ x_2 & y_2 \\ x_3 & y_3 \end{bmatrix}^e$$

or simply

$$\mathbf{J}^e = \begin{bmatrix} (x_2 - x_1) & (y_2 - y_1) \\ (x_3 - x_1) & (y_3 - y_1) \end{bmatrix}^e \tag{11.28}$$

which is also constant. The determinant of this 2 by 2 matrix is $|\mathbf{J}^e| = (x_2 - x_1)^e(y_3 - y_1)^e - (x_3 - x_1)^e(y_2 - y_1)^e = 2A^e$. In general the inverse Jacobian in two-dimensions is

$$\mathbf{J}^{-1} = \frac{1}{|\mathbf{J}|} \begin{bmatrix} \partial y/\partial s & -\partial y/\partial r \\ -\partial x/\partial s & \partial x/\partial r \end{bmatrix} .$$

For the three node triangle this is simply

$$\mathbf{J}^{e^{-1}} = \frac{1}{2A^e} \begin{bmatrix} (y_3 - y_1) & -(y_2 - y_1) \\ -(x_3 - x_1) & (x_2 - x_1) \end{bmatrix}^e . \tag{11.29}$$

For most other elements it is common to form these quantities numerically by utilizing the numerical values of $\mathbf{R}^e$ given in the data.

The use of the local coordinates in effect represents a change of variables. In this sense the Jacobian has another important function. The determinant of the Jacobian, $|\mathbf{J}|$, relates differential changes in the two coordinate systems. That is,

$$d\ell = dx = |\mathbf{J}| \; dr$$

$$da = dx\,dy = |\mathbf{J}| \; dr\,ds$$

$$dv = dx\,dy\,dz = |\mathbf{J}| \; dr\,ds\,dt \tag{11.30}$$

in one-, two-, and three-dimensional problems. This becomes important when integrals are considered.

The integral definitions of the element matrices usually involve the global derivatives of the quantity of interest. From Eq. (11.23) it is seen that the local derivatives of V are related to the nodal parameters by

$$\begin{Bmatrix} \frac{\partial V}{\partial r} \\ \frac{\partial V}{\partial s} \\ \frac{\partial V}{\partial t} \end{Bmatrix} = \begin{bmatrix} \frac{\partial}{\partial r} \mathbf{H} \\ \frac{\partial}{\partial s} \mathbf{H} \\ \frac{\partial}{\partial t} \mathbf{H} \end{bmatrix} \mathbf{V}^e,$$

or, symbolically,

$$\boldsymbol{\partial}_{\ell} V(r,s,t) = \boldsymbol{\Delta}(r,s,t)\mathbf{V}^e. \tag{11.31}$$

To relate the global derivatives of V to the nodal parameters, $\mathbf{V}^e$, one substitutes the above expression into Eq. (11.25) to obtain

$$\boldsymbol{\partial}_g V = \mathbf{J}^{-1}\boldsymbol{\Delta}\mathbf{V}^e,$$

or

$$\boldsymbol{\partial}_g V(r,s,t) = \mathbf{d}(r,s,t)\mathbf{V}^e;$$

where

$$\mathbf{d}(r,s,t) = \mathbf{J}(r,s,t)^{-1}\boldsymbol{\Delta}(r,s,t). \tag{11.32}$$

The matrix **d** is very important since it relates the global derivatives of the quantity of interest to the quantity's nodal values. For the sake of completeness, note that **d** can be partitioned as

$$\mathbf{d}(r,s,t) = \begin{bmatrix} d_x \\ \hline d_y \\ \hline d_z \end{bmatrix} = \begin{bmatrix} \frac{\partial}{\partial x} \mathbf{H} \\ \hline \frac{\partial}{\partial y} \mathbf{H} \\ \hline \frac{\partial}{\partial z} \mathbf{H} \end{bmatrix} = \boldsymbol{\partial}_g \mathbf{H} \tag{11.33}$$

so that each row represents a derivative of the interpolation functions with respect to a global coordinate direction. In practice the **d** matrix usually exists only in numerical form at selected points. The three node triangle is an exception since **J**, **Δ**, and thus **d** are all constant. Substituting the results from Eqs. (11.27) and (11.29) into (11.31) yields

$$\mathbf{d}^e = \frac{1}{2A^e} \begin{bmatrix} (y_2 - y_3) & (y_3 - y_1) & (y_1 - y_2) \\ (x_3 - x_2) & (x_1 - x_3) & (x_2 - x_1) \end{bmatrix}^e . \tag{11.34}$$

Any finite element analysis ultimately leads to the evaluation of the integrals that define the element and/or boundary segment matrices. The element matrices, $\mathbf{S}^e$ or $\mathbf{C}^e$, are usually defined by integrals of the symbolic form

$$\mathbf{A}^e = \int\int\int_{\Omega^e} \mathbf{B}^e(x,y,z)\ dx\ dy\ dz,$$

where $\mathbf{B}^e$ is usually the sum of products of other matrices involving the element interpolation functions, **H**, their derivatives, **d**, and problem properties. With the element formulated in terms of the local (r,s,t) coordinates, where $\mathbf{B}^e(x,y,z)$ is transformed into $\tilde{\mathbf{B}}^e(r,s,t)$, this expression must be rewritten in local coordinates as

$$\mathbf{A}^e = \int_{-1}^{1}\int_{-1}^{1}\int_{-1}^{1} \tilde{\mathbf{B}}^e(r,s,t)\ \left|\ \mathbf{J}^e(r,s,t)\right|\ dr\ ds\ dt. \tag{11.35}$$

Note that Eq. (11.30) has been used in the above expression. In practice, one would often use numerical integration (see Chapter 6 and Section 11.7) to obtain

$$\mathbf{A}^e = \sum_{i=1}^{NIP} W_i \tilde{\mathbf{B}}^e(r_i,s_i,t_i)\ \left|\ \mathbf{J}^e(r_i,s_i,t_i)\right| \tag{11.36}$$

where $\tilde{\mathbf{B}}^e$ and $|\mathbf{J}|$ are evaluated at each of the NIP integration points and where $(r_i,\ s_i,\ r_i)$ and W_i denote the tabulated abscissae and weights, respectively. These concepts will be considered in more detail later in this chapter.

11.5 Exact Integration in 2-D and 3-D

Recall that the finite element analysis techniques are always based on an integral formulation. At the very minimum it will always be necessary to integrate at least an element square matrix. This means that every coefficient function in the matrix must be integrated. In the following sections various methods will be considered for evaluating the typical integrals that arise.

11.5.1 Global Coordinate Integration

Most simple finite element matrices for two-dimensional problems are based on the use of linear triangular or quadrilateral elements. Since a quadrilateral can be divided into two or more triangles, only exact integrals over arbitrary triangles will be considered here. Integrals over triangular elements commonly involve integrands of the form

$$I = \int_A x^m y^n dxdy \qquad (11.37)$$

where A is the area of a typical triangle. When $0 \leq (m + n) \leq 2$, the above integral can easily be expressed in closed form in terms of the spatial coordinates of the three corner points. For a right-handed coordinate system, the corners must be numbered in counter-clockwise order. In this case, the above integrals are given in Table XI.I. These integrals should be recognized as the area, and first and second moments of the area. If one had a volume of revolution that had a triangular cross-section in the ρ-z plane, then one should recall that

$$I = \int_V \rho f(\rho,z) d\rho dz d\phi = 2\pi \int_A \rho f(\rho,z) d\rho dz$$

so that similar expressions could be used to evaluate the volume integrals. The above closed form integrals are included in subroutine TRGEOM, which is shown in Fig. 11.11. One enters this routine with the values of the integers m and n, the corner coordinates, and returns with the desired area (or volume) integral. Similar operations for quadrilaterals are performed by splitting the quadrilateral into two triangles and making two calls to TRGEOM to evaluate the integral. The exact integration over a tetrahedron can also be written in closed form. Typical equations for that purpose are given in Reference [24].

11.5.2 Unit Coordinate Integration

The utilization of global coordinate interpolation is becoming increasingly rare. However, as we have seen, the use of non-dimensional local coordinates is common. Thus we often see local coordinate polynomials integrated over the physical domain of an element. Section 6.3 presented some typical unit coordinate integrals in 1-D, written in exact closed form. These concepts can be extended to two- and three-dimensional elements. For example, consider an integration over a triangular element. It is known that for an element with a constant Jacobian

$$I = \int_A r^m s^n da = \frac{2A\Gamma(m + 1)\ \Gamma(n + 1)}{\Gamma(3 + m + n)}$$

where Γ denotes the Gamma function. Restricting consideration to positive integer values of the exponents, m and n, yields

$$I = 2A^e \frac{m!\ n!}{(2 + m + n)!} = A^e/K_{mn}, \tag{11.38}$$

where ! denotes the factorial and K_{mn} is an integer constant given in Table XI.II for common values of m and n. Similarly for the tetrahedron element

$$I^e = \int_{V^e} r^m\ s^n\ t^p\ dv = 6V^e\ \frac{m!\ n!\ p!}{(3 + m + n + p)!}$$

Thus, one notes that common integrals of this type can be evaluated by simply multiplying the element characteristic (i.e., global length, area, or volume) by known constants which could be stored in a data statement.

To illustrate the application of these equations in evaluating element matrices, we consider the following example for the three node triangle:

Table XI.I

Exact Integrals For A Triangle

m	n	I
0	0	$\int dA = 1/2\ (x_1(y_2-y_3) + x_2(y_3-y_1) + x_3(y_1-y_2)) = A$
0	1	$\int y dA = A\bar{y} = A(y_1 + y_2 + y_3)/3$
1	0	$\int x dA = A\bar{x} = A(x_1 + x_2 + x_3)/3$
0	2	$\int y^2 dA = 1/12\ A({y_1}^2 + {y_2}^2 + {y_3}^2 + 9\bar{y}^2)$
1	1	$\int xy dA = 1/12\ A(x_1y_1 + x_2y_2 + x_3y_3 + 9\bar{x}\ \bar{y})$
2	0	$\int x^2 dA = 1/12\ A({x_1}^2 + {x_2}^2 + {x_3}^2 + 9\bar{x}^2)$

Table XI.II Exact Area Integrals in Unit Coordinates

$$I = \int_A r^m s^n \, da = A/K$$

m	0	1	0	2	1	0	3	2	1	0	4	3	2	1	0
n	0	0	1	0	1	2	0	1	2	3	0	1	2	3	4
K	1	3	3	6	12	6	10	30	30	10	15	60	90	60	15

$$I = \int_{A^e} \mathbf{H}^T da = \int_{A^e} \begin{Bmatrix} (1-r-s) \\ r \\ s \end{Bmatrix} da = \begin{Bmatrix} A^e - A^e/3 - A^e/3 \\ A^e/3 \\ A^e/3 \end{Bmatrix}$$

$$= \frac{A^e}{3} \begin{Bmatrix} 1 \\ 1 \\ 1 \end{Bmatrix} .$$

11.5.3 Simplex Coordinate Integration

A simplex region is one where the minimum number of vertices is one more than the dimension of the space. These were illustrated in Fig. 3.2. Some analysts like to define a set of **simplex coordinates.** If there are N nodes then N non-dimensional coordinates, L_i, $1 \leq i \leq N$, are defined and constrained so that

$$1 = \sum_{i=1}^{N} L_i$$

at any point in space. Thus they are not independent. However they can be used to simplify certain recursion relations. In physical spaces these coordinates are sometimes called line coordinates, **area coordinates,** and **volume coordinates.**

At a given point in the region we can define the simplex coordinate for node j, L_j, in a generalized manner. It is the ratio of the generalized volume from the point to all other vertices (other than j) and the total generalized volume of the simplex. This is illustrated in Fig. 11.12. If the simplex has a constant Jacobian (e.g. straight sides) then the exact form of the integrals of the simplex coordinates are simple. They are

$$\int_{\ell} L_1^{a} L_2^{b} \, d\ell = \frac{a!\ b!}{(a + b + 1)!} \ (\ell)$$

$$\int_{A} L_1^{a} L_2^{b} L_3^{c} \, da = \frac{a!\ b!\ c!}{(a + b + c + 2)!} \ (2A)$$

$$\int_{V} L_1^{a} L_2^{b} L_3^{c} L_4^{d} \, dv = \frac{a!\ b!\ c!\ d!}{(a + b + c + d + 3)!} \ (6V).$$

```
      SUBROUTINE  TRGEOM ( NCODE, COORD, VALUE )
C     GEOMETRIC PROPERTIES OF AN ARBITRARY TRIANGLE
      DIMENSION  COORD(3,2)
C     VALUE = INTEGRAL (X**M)*(Y**N)*DA   (0<=M,N<=2)
C     VALUE BY NCODE: 1-AREA, 2-VOL REV ABOUT X,
C     3-VOL REV ABOUT Y, 4-FIRST MOMENT ABOUT X,
C     5-FIRST MOMENT ABOUT Y, 6-SEC MOMENT ABOUT X,
C     7-SEC MOMENT WRT X-Y AXES, 8-SEC MOMENT ABOUT Y
      PARAMETER ( TWOPI = 6.2831853 )
      XI = COORD(1,1)
      XJ = COORD(2,1)
      XK = COORD(3,1)
      YI = COORD(1,2)
      YJ = COORD(2,2)
      YK = COORD(3,2)
      AREA = 0.5*(XI*(YJ-YK)+XJ*(YK-YI)+XK*(YI-YJ))
      XB = (XI+XJ+XK)/3.0
      YB = (YI+YJ+YK)/3.0
      GO TO (10,20,30,40,50,60,70,80), NCODE
   10 VALUE = AREA
      RETURN
   20 VALUE = TWOPI*AREA*YB
      IF ((ABS(YI)+ABS(YJ)+ABS(YK)) .NE. ABS(YI+YJ+YK))
     1   WRITE (6 ,5000)
 5000 FORMAT(' INVALID DATA FOR VOLUME OF REVOLUTION')
      RETURN
   30 VALUE = TWOPI*AREA*XB
      IF ((ABS(XI)+ABS(XJ)+ABS(XK)) .NE. ABS(XI+XJ+XK))
     1   WRITE (6 ,5000)
      RETURN
   40 VALUE = AREA*YB
      RETURN
   50 VALUE = AREA*XB
      RETURN
   60 VALUE = AREA*(YI*YI+YJ*YJ+YK*YK+9.0*YB*YB)/12.0
      RETURN
   70 VALUE = AREA*(XI*YI+XJ*YJ+XK*YK+9.0*XB*YB)/12.0
      RETURN
   80 VALUE = AREA*(XI*XI+XJ*XJ+XK*XK+9.0*XB*XB)/12.0
      RETURN
      END
```

Fig. 11.11 Subroutine TRGEOM For Exact Integrals

(a) One-dimensional

$L_1 = \dfrac{\text{length } P2}{\text{length } 12}$

$1 = L_1 + L_2$

(b) Two-dimensional

$L_1 = \dfrac{\text{Area } P23}{\text{Area } 123}$

$L_2 = \dfrac{\text{Area } P13}{\text{Area } 123}$

$1 = L_1 + L_2 + L_3$

(c) Three-dimensional

$L_1 = \dfrac{\text{Volume } P234}{\text{Volume } 1234}$

$L_2 = \dfrac{\text{Volume } P134}{\text{Volume } 1234}$

$L_3 = \dfrac{\text{Volume } P124}{\text{Volume } 1234}$

$1 = L_1 + L_2 + L_3 + L_4$

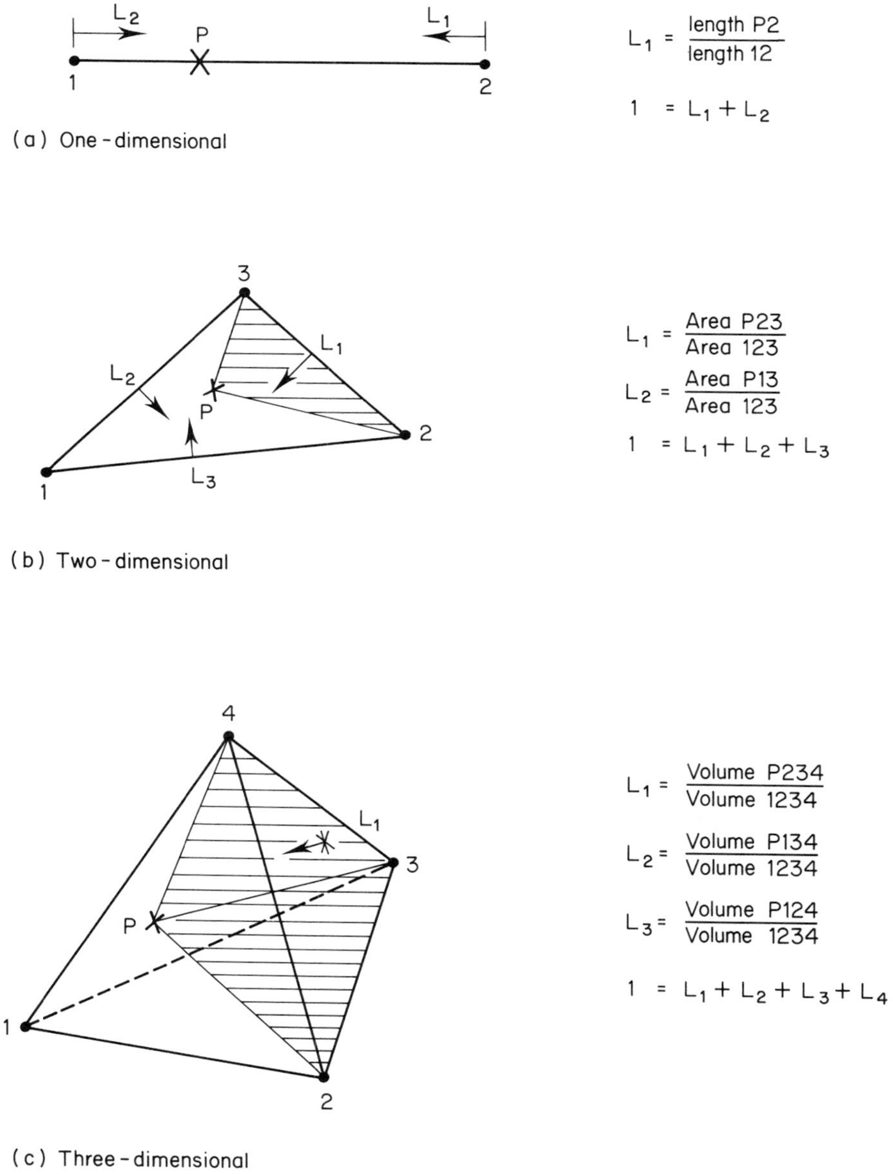

Fig. 11.12 Simplex Coordinates

11.6 Numerical Integration in 2-D and 3-D

In many cases it is impossible or impractical to integrate the expression in closed form and numerical integration must therefore be utilized. If one is using sophisticated elements, it is almost always necessary to use numerical integration. Similarly, if the application is complicated, e.g. the solution of a non-linear ordinary differential equation, then even simple one-dimensional elements can require numerical integration. Many analysts have found that the use of numerical integration simplifies the programming of the element matrices. This results from the fact that lengthy algebraic expressions are avoided and thus the chance of algebraic and/or programming errors is reduced. There are many numerical integration methods available. Only those methods commonly used in finite element applications will be considered here.

11.6.1 Unit Coordinate Quadrature

Numerical quadrature in one-dimension was introduced in Sec. 6.4. There we saw that an integral is replaced with a summation of functions evaluated at tabulated points and then multiplied by tabulated weights. The same procedure applies to all numerical integration rules. The main difficulty is to obtain the tabulated data. For triangular unit coordinate regions the weights, W_i, and abscissae (r_i,s_i) are less well known. Several different rules on the unit triangle are tabulated in Figs. 11.13 and 11.14. Figure 11.13 presents rules that yield points that are symmetric with respect to all corners of the triangle. Figure 11.14 presents a rule that yields points that are symmetric with respect to the line $r = s$. They are known as Radau rules. Similar data are tabulated in other references such as [1].

As before, one approximates an integral of $f(x,y) = F(r,s)$ over a triangle by

$$I = \int f(x,y)dxdy = \sum_{i=1}^{n} W_i F(r_i,s_i)\ |J_i|.$$

As a simple example of integration over a triangle, let $f = y$ and consider the integral over a triangle with its three vertices at (0,0), (3,0) and (0,6), respectively, in (x,y) coordinates. Then the area $A = 9$ and the Jacobian is a constant $|J| = 18$. For a three point quadrature rule the integral is thus given by

$$I = \sum_{i=1}^{3} W_i y_i |J_i| .$$

Since our interpolation defines

$$y(r,s) = y_1 + (y_2 - y_1)r + (y_3 - y_1)s$$
$$= 0 + 0 + 6s = 6s$$

the transformed integran is $F(r,s) = 6s$. Thus at integration point, i, $F(r_i,s_i) = 6s_i$. Substituting the three points from Fig. 11.13 and factoring out the constant Jacobian gives

$$I = 18[(6(1/6))(1/6) + (6(1/6))(1/6) + (6(2/3))(1/6)] = 18$$

which is the exact solution.

Table XI.III gives a tabulation of symmetric quadrature rules over the unit triangle. Additional unsymmetric rules are found in [1] and elsewhere.

11.6.2 **Natural Coordinate Quadrature**

Here we assume that the coordinates are in the range of -1 to +1. In this space it is common to employ Gaussian quadratures. The one-dimensional rules were discussed in Sec. 6.4. For a higher number of space dimensions one obtains a multiple summation for evaluating the integral. For example, a typical integration in two dimensions

$$I = \int_{-1}^{1} \int_{-1}^{1} f(r,s)drds$$

becomes

$$I \approx \sum_{j=1}^{n} \sum_{k=1}^{n} f(r_j, s_k)W_j W_k$$

for n integration points in each dimension. This can be written as a single summation as

$$I \approx \sum_{i=1}^{m} f(r_i, s_i)W_i$$

Table XI.III

Symmetric Quadrature Data for the Unit Triangle:

$$\int_0^1 \int_0^{1-r} f(r, s)\, dr\, ds = \sum_{i=1}^{n} f(r_i, s_i) W_i$$

n	P*	i	r_i	s_i	W_i
1	1	1	1/3	1/3	1/2
3	2	1	1/6	1/6	1/6
		2	2/3	1/6	1/6
		3	1/6	2/3	1/6
3	2	1	1/2	0	1/6
		2	1/2	1/2	1/6
		3	0	1/2	1/6
4	3	1	1/3	1/3	-9/32
		2	3/5	1/5	25/96
		3	1/5	3/5	25/96
		4	1/5	1/5	25/96
4	3	1	1/3	1/3	-9/32
		2	11/15	2/15	25/96
		3	2/15	2/15	25/96
		4	2/15	11/15	25/96
7	4	1	0	0	1/40
		2	1/2	0	1/15
		3	1	0	1/40
		4	1/2	1/2	1/15
		5	0	1	1/40
		6	0	1/2	1/15
		7	1/3	1/3	9/40

* P = Degree of Polynomial for exact integration

Table XI.IV Weights and Abscissae for Natural Coordinate Gaussian Quadrature On a Quadrilateral

$$\int_{-1}^{1}\int_{-1}^{1} f(r,s)\,dr\,ds = \sum_{i=1}^{n} f(r_i,s_i)W_i$$

n	i	r_i	s_i	W_i
1	1	0	0	4
4	1	$-1/\sqrt{3}$	$-1/\sqrt{3}$	1
	2	$+1/\sqrt{3}$	$-1/\sqrt{3}$	1
	3	$-1/\sqrt{3}$	$+1/\sqrt{3}$	1
	4	$+1/\sqrt{3}$	$+1/\sqrt{3}$	1
9	1	$-\sqrt{3/5}$	$-\sqrt{3/5}$	25/81
	2	0	$-\sqrt{3/5}$	40/81
	3	$+\sqrt{3/5}$	$-\sqrt{3/5}$	25/81
	4	$-\sqrt{3/5}$	0	40/81
	5	0	0	64/81
	6	$+\sqrt{3/5}$	0	40/81
	7	$-\sqrt{3/5}$	$+\sqrt{3/5}$	25/81
	8	0	$+\sqrt{3/5}$	40/81
	9	$+\sqrt{3/5}$	$+\sqrt{3/5}$	25/81

where $m = n^2$, $i = j + (k - 1)n$, and where $r_i = \alpha_j$, $s_i = \alpha_k$, and $W_i = W_j W_k$. Here α_j and W_j denote the tabulated one-dimensional absiccae and weights given in Sec. 6.4. A similar rule can be given for a three-dimensional region [1]. The result of the above summation is given in Table XI.IV.

11.7 Typical Source Distribution Integrals

Previously we introduced the contributions of distributed source terms through integrals such as Eqs. (4.32), (4.65), and (4.66). For the C° continuity line elements we had

$$\mathbf{C}_Q^e = \int_{\ell^e} \mathbf{H}^{e^T} Q^e dx.$$

Similar forms occur in two-dimensional problems (like Eqs. (12.10) or (13.22)). Then typically one has

$$\mathbf{C}_Q^e = \int_{A^e} \mathbf{H}^{e^T} Q^e da.$$

If the typical source or forcing term, Q^e, varies with position we usually use the interpolation functions to define it in terms of the nodal values, $\mathbf{Q}^e$, as

$$Q^e = \mathbf{H}^e \mathbf{Q}^e. \tag{11.39}$$

Thus a common element (or boundary) integral for the consistent nodal sources is

$$\mathbf{C}_Q^e = \int \mathbf{H}^{e^T} \mathbf{H}^e dx\ \mathbf{Q}^e. \tag{11.40}$$

The previous sections present analytic and numerical methods for evaluating these integrals. Figure 11.15 shows the typical analytic results for the two and three node line integrals. For linear or constant source distributions the normalized nodal resultants are summarized in Fig. 11.16. Once one goes beyond the linear (two-node) element the consistent results usually differ from physical intuition estimates. Thus you must rely on the mathematics or the summaries in the above figures.

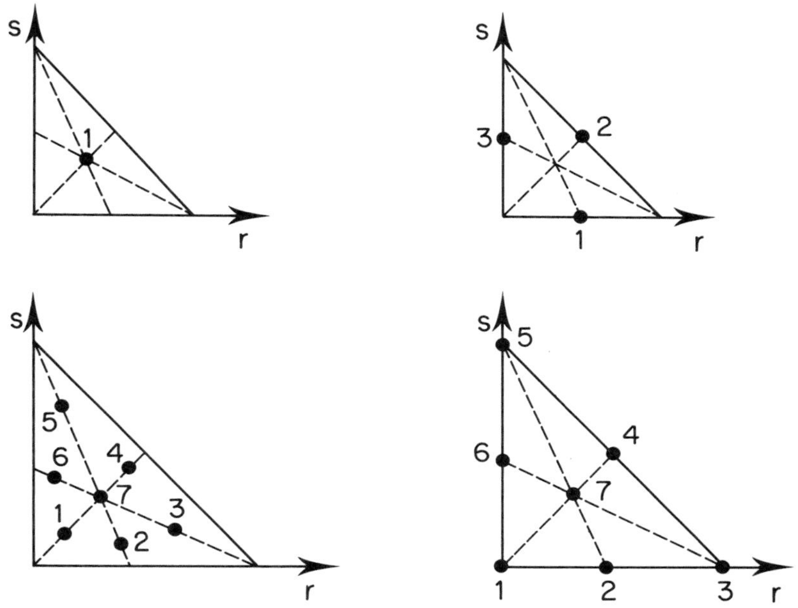

Fig. 11.13 Symmetric Quadriture Points for Triangles

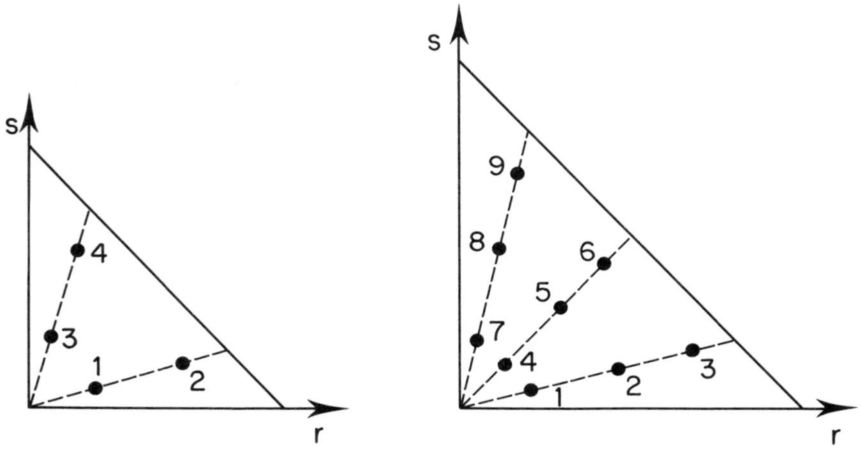

Fig. 11.14 Non-Symmetric Quadrature Points

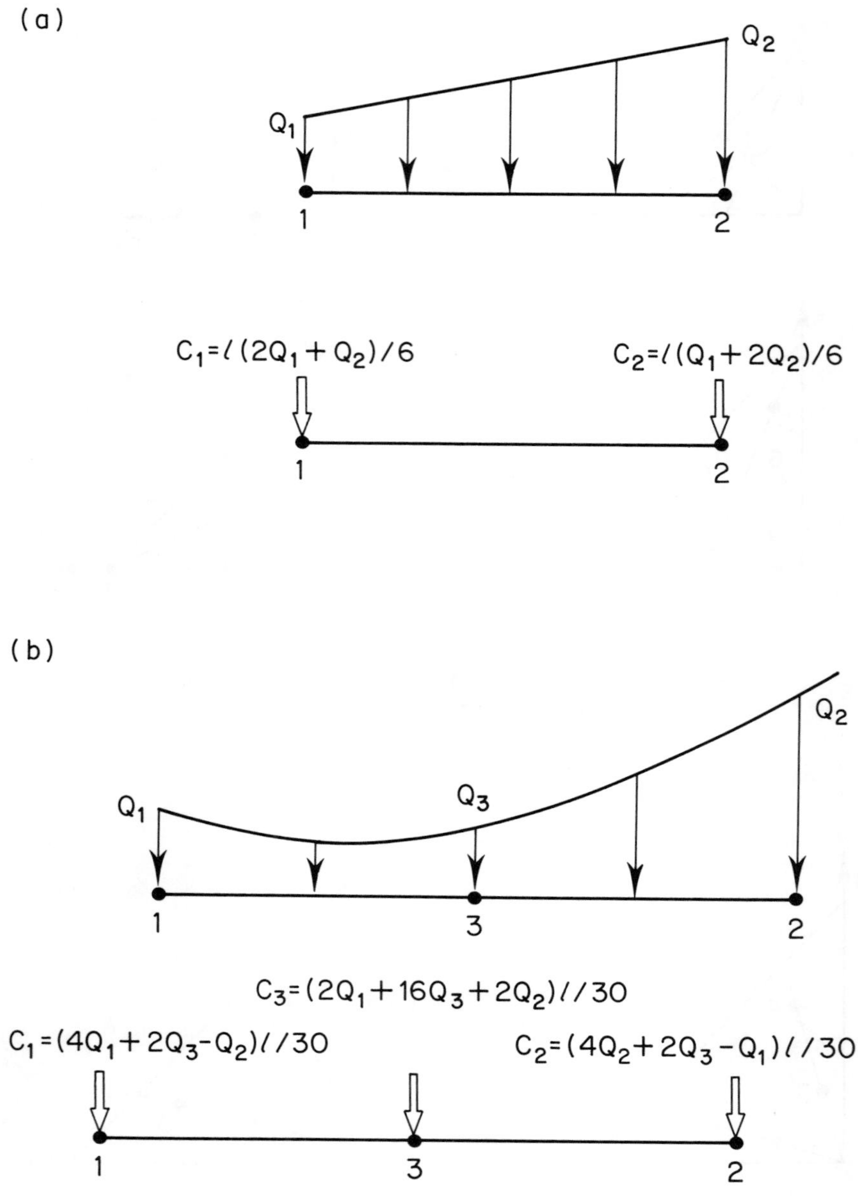

Fig. 11.15 General Consistent Line Sources

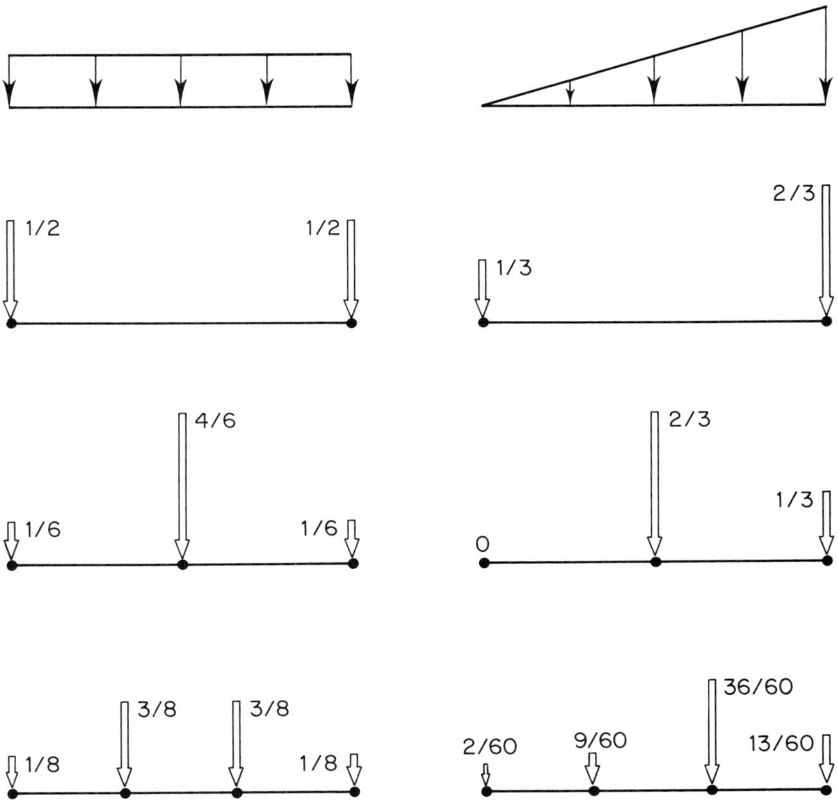

Fig. 11.16 Consistent Distributions for a Unit Source

Many programs will numerically integrate the source distributions for any element shape. If the source acts on an area shaped like the parent element (constant Jacobian) then we can again easily evaluate the integrals analytically. For a uniform source over an area the consistent nodal contributions for quadrilaterals and triangles are shown in Figs. 11.17 and 11.18, respectively. Note that the Serendipity families can actually develop negative contributions. Triangular and Lagrangian elements do not have that behavior for uniform sources.

Sometimes manual input of these nodal resultants is necessary. Then one may wish to use reasonable engineering approximations to reduce the labor involved. Figure 11.19 illustrates a typical set of resultants that could be used in such cases. If the source varies linearly from zero then a more complicated set of nodal resultants are obtained as shown in Figs. 11.20 and 11.21. Of course, a general linear loading can be formed by combining the resultants for the constant and triangular sources.

11.8 Exercises

1. Consider an eight node quadrilateral element as in Fig. 11.3 or Eqs. (11.19) and (11.20). Evaluate the eight shape functions for b = -1. Next assume that the physical coordinates of the nodes are such that the a = 0 node is located in the center between the two end nodes (e.g., $x(0) = (x(+1) + x(-1))/2$). Determine the shape of the edge in physical space.
2. The interpolation functions for the linear triangle are given in physical coordinates in Eqs. (11.13) to (11.15). a) Verify that the sum of the interpolation functions is unity at any point (x,y). b) Verify that the sum of the global derivatives are identically zero at any point (x,y). c) Use calculus to verify that the area integral of an interpolation function is A/3.
3. Algebraically evaluate the linear triangle interpolation functions at each of the three nodes. Sketch the spatial variation of each function over the element.
4. Use the summations in Sec. 11.6.2 to verify the data in Table XI.IV for: a) the 4 point rule, b) 9 point rule.
5. The triangular load on the six node triangle in Fig. 11.21 can be combined three times to model a uniform source. Do this and compare the resultant with that shown in Fig. 11.18.

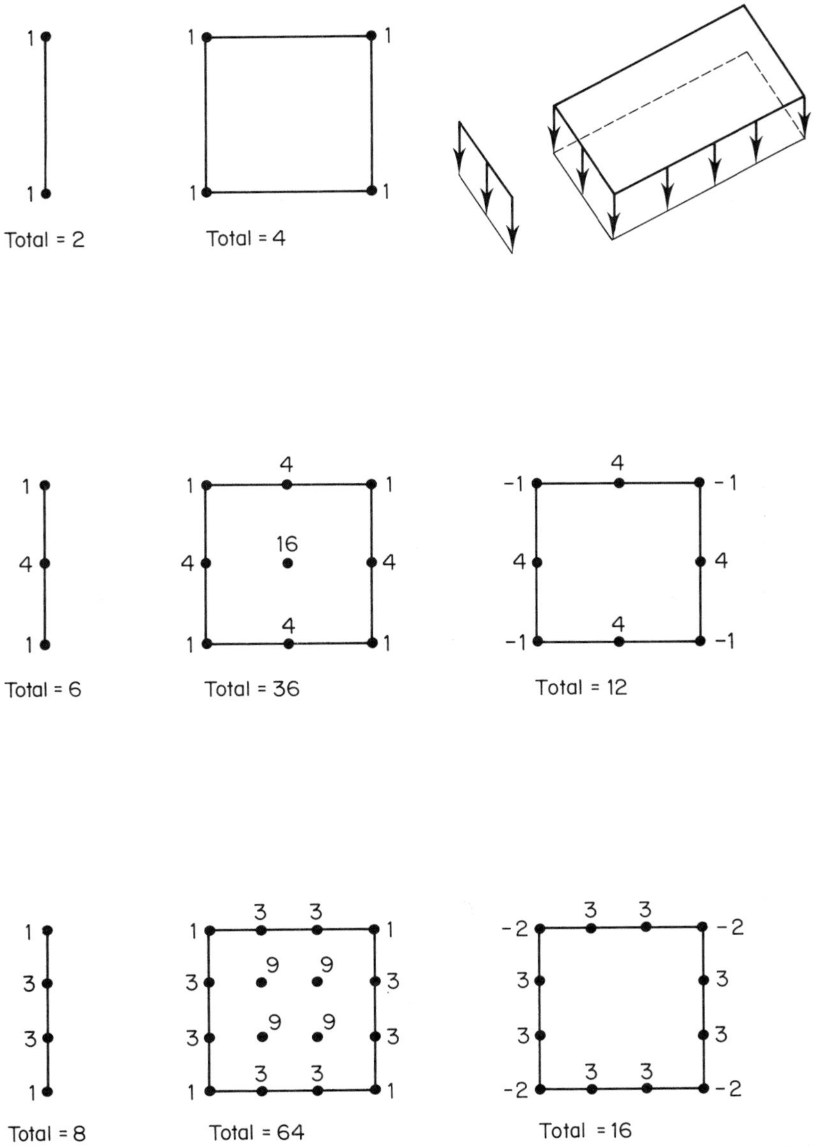

Fig. 11.17 Nodal Resultants for a Uniform Source on a Quadrilateral

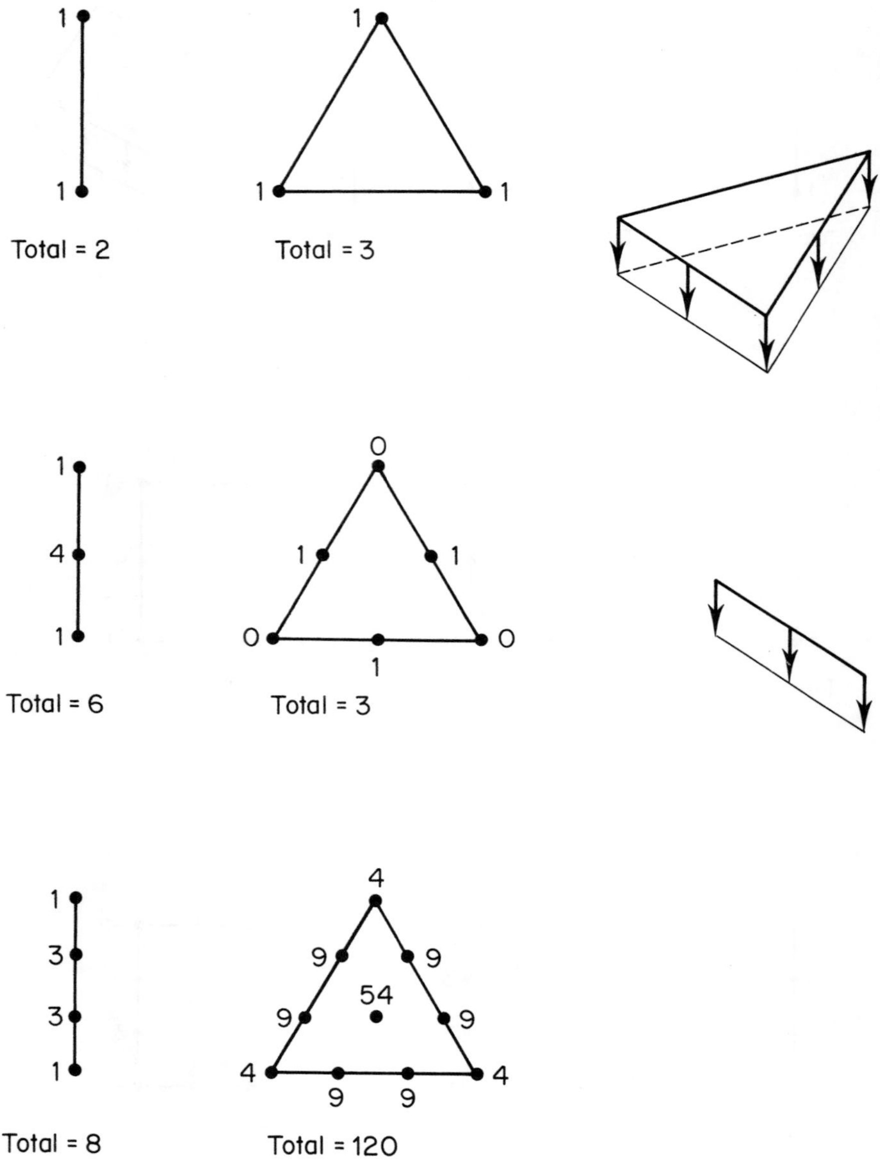

Fig. 11.18 Nodal Resultants for a Uniform Source on a Triangle

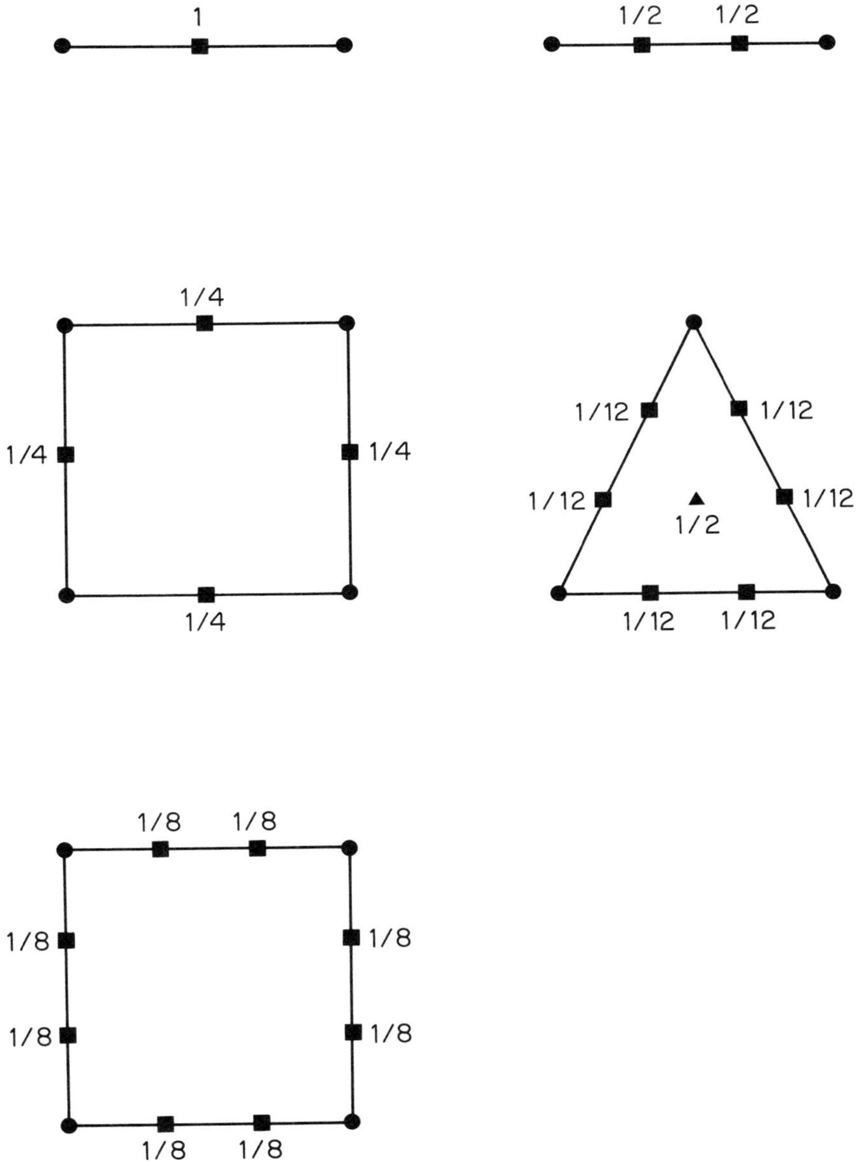

Fig. 11.19 The Complete Polynomial Forms

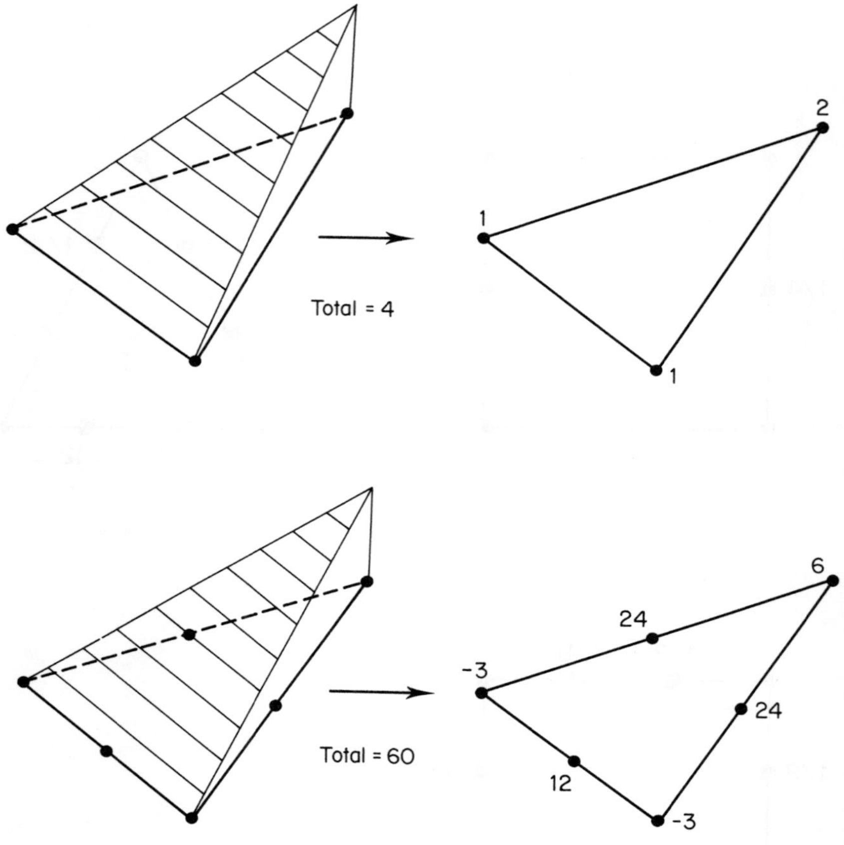

Fig. 11.20 Nodal Resultants for Linear Source Distribution on a Triangle

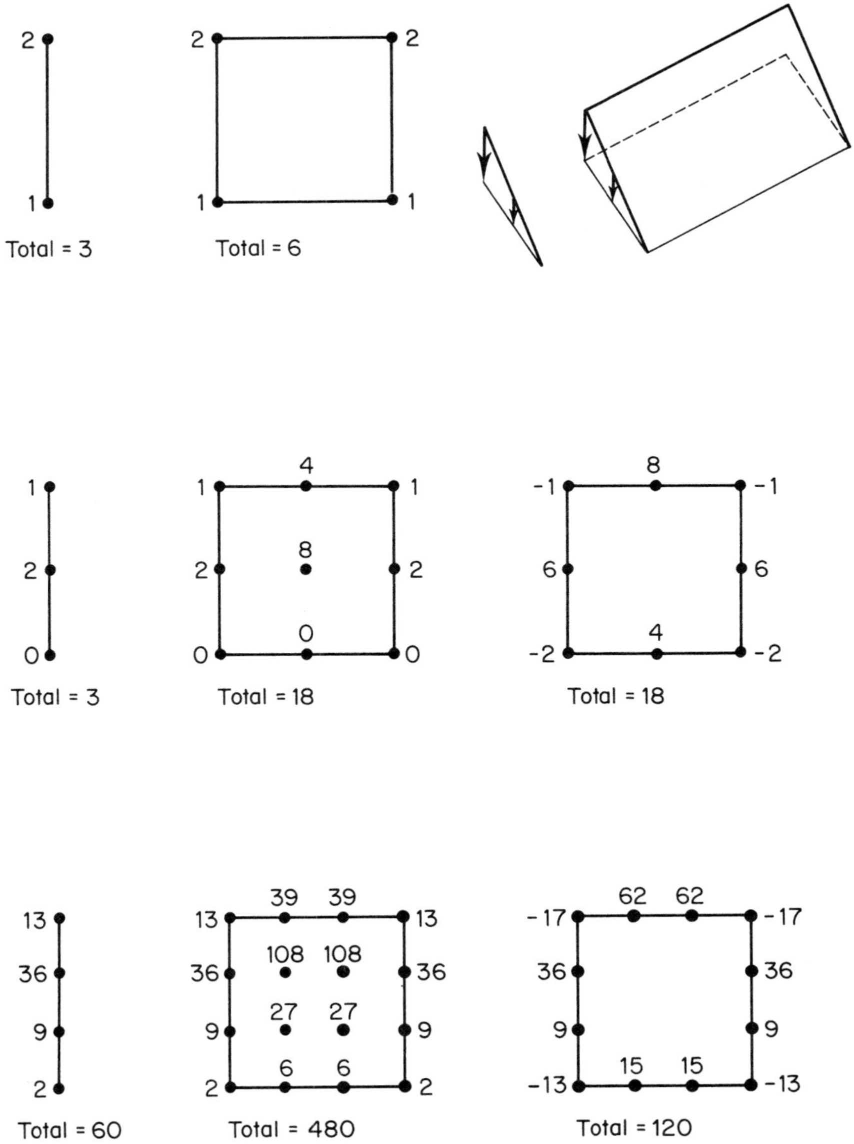

Fig. 11.21 Nodal Resultants for a Linear Source on a Quadrilateral

6. For the nine node quadrilateral, use two combinations of the resultant in Fig. 11.20 to find the resultant for a uniform source. Compare against the values in Fig. 11.17.
7. Use local unit coordinates and integrate a linear source on the six node triangle to verify Fig. 11.21.

12. TWO-DIMENSIONAL HEAT TRANSFER

12.1 Introduction

Earlier it was demonstrated that the basic relationships for the finite element formulation of elasticity problems could be obtained simply by minimizing the total potential energy of the system without any direct reference to the usual static equilibrium equations. In many areas of engineering and physics it is possible to obtain "exact" solutions to problems by minimizing some functional, subjected to certain boundary conditions. In the case of elasticity this functional was physically interpreted as the total potential energy of the system. For many areas of engineering and physics the functional may simply be a mathematically defined quantity. In other fields the physical interpretation may not be obvious. For example, in ideal fluid flow the functional may represent the rate of entropy production.

The physical behavior governing a variety of problems in engineering can be described by the well known Laplace and Poisson differential equations. The analytic solution of these equations in two- and three-dimensional field problems can present a formidable task, especially in the case where there are complex boundary conditions and irregularly shaped regions. The finite element formulation of this class of problem by using variational methods has proven to be a very effective and versatile approach to the solution. Previous difficulties associated with irregular geometry and complex boundary conditions are virtually eliminated.

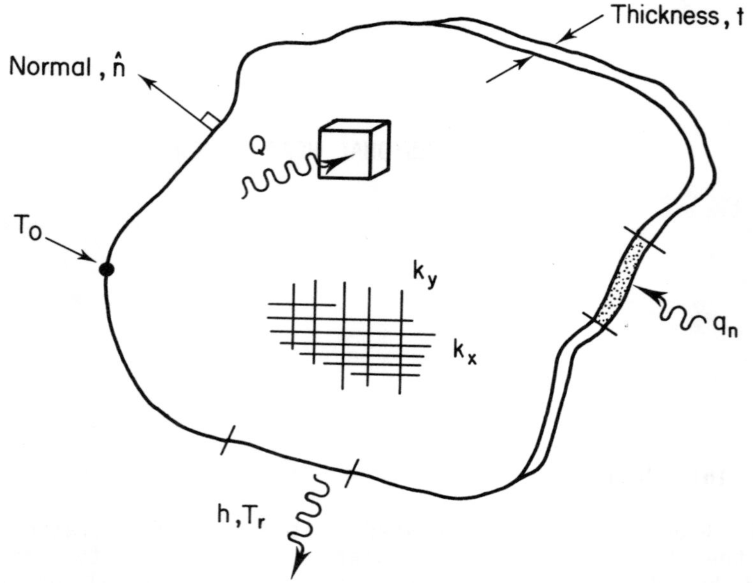

Fig. 12.1 An anisotropic conduction region

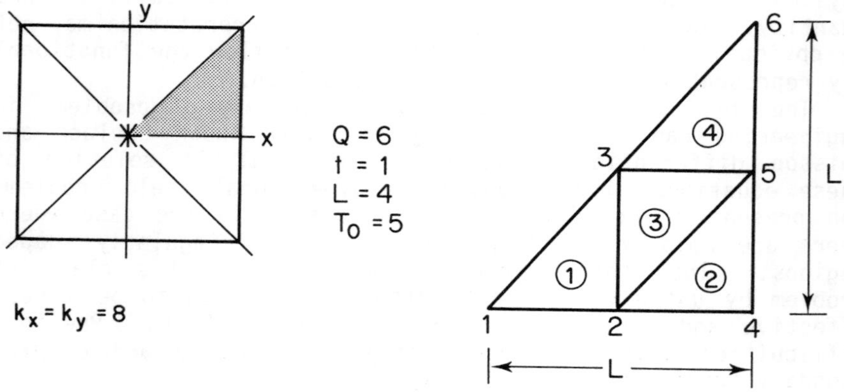

Fig. 12.2 A one-eighth symmetry model of a square

a. the temperature, T, is specified on the boundary, i.e.

$$T = T(s) \text{ on } \Gamma_s. \tag{12.3}$$

b. boundary heat flux is specified at a point, i.e.

$$k_x \frac{\partial T}{\partial x} \ell_x + k_y \frac{\partial T}{\partial y} \ell_y + q + h(T - T_r) = 0 \text{ on } \Gamma_q$$

or

$$k_n \frac{\partial T}{\partial n} + q + h(T - T_r) = 0, \tag{12.4}$$

where ℓ_x and ℓ_y are the direction cosines of the outward normal to the boundary surface, q represents the heat flux per unit of surface, and $h(T - T_r)$ is the convection heat loss.

As stated previously an alternative formulation to the above heat conduction problem is possible using the calculus of variations. Euler's theorem of the calculus of variations states that if the integral

$$I(u) = \int_V f(x, y, z, u, \frac{\partial u}{\partial x}, \frac{\partial u}{\partial y}, \frac{\partial u}{\partial z}) \, dx \, dy \, dz + \int_S (qu + h\, u^2/2) \, da \tag{12.5}$$

is to be minimized, the necessary and sufficient condition for this minimum to be reached is that the unknown function u(x,y,z) satisfy the following differential equation

$$\frac{\partial}{\partial x} \frac{\partial f}{\partial(\partial u/\partial x)} + \frac{\partial}{\partial y} \frac{\partial f}{\partial(\partial u/\partial y)} + \frac{\partial}{\partial z} \frac{\partial f}{\partial(\partial u/\partial z)} - \frac{\partial f}{\partial u} = 0 \tag{12.6}$$

within the region, provided u satisfies the essential boundary conditions in both cases. We can verify that the minimization of the volume integral

$$I = \int_V [\frac{1}{2} \{k_x (\frac{\partial T}{\partial x})^2 + k_y (\frac{\partial T}{\partial y})^2 + k_z (\frac{\partial T}{\partial z})^2\} - QT] \, dx \, dy \, dz + \int_S [qT + hT^2/2] \, da \tag{12.7}$$

Some examples of problems frequently encountered in engineering practice falling into this category are: heat conduction, seepage through porous media, torsion of prismatic shafts, irrotational flow of ideal fluids, distribution of magnetic potential, etc. The following development will be concerned with the details of formulating the solution to the two-dimensional, steady-state heat conduction problem. The approach is general however, and by redefining the physical quantities involved the formulation is equally applicable to other problems involving the Poisson equation.

12.2 **Variational Formulation**

We can obtain from any book on heat transfer the governing differential equation for steady and un-steady state heat conduction. The most general form of the heat conduction equation is the transient three-dimensional equation:

$$\frac{\partial}{\partial x}\left(k_x \frac{\partial T}{\partial x}\right) + \frac{\partial}{\partial y}\left(k_y \frac{\partial T}{\partial y}\right) + \frac{\partial}{\partial z}\left(k_z \frac{\partial T}{\partial z}\right) + Q = \frac{\partial}{\partial t}(\rho c T) \qquad (12.1)$$

where,

k_x, k_y, k_z = thermal conductivity coefficients

T = unknown temperature distribution
Q = heat generation per unit volume
ρ = density
c = heat capacity.

If we focus our attention to the two-dimensional ($\partial/\partial z = 0$) steady-state ($\partial/\partial t = 0$) problem, such as Fig. 12.1, the governing equation becomes

$$\frac{\partial}{\partial x}\left(k_x \frac{\partial T}{\partial x}\right) + \frac{\partial}{\partial y}\left(k_y \frac{\partial T}{\partial y}\right) + Q = 0 \qquad (12.2)$$

in which k_x, k_y, and Q are known. Equations (12.1) or (12.2) together with the appropriate boundary conditions specify the problem completely. The two most commonly encountered boundary conditions are those in which

leads directly to the formulation equivalent to Eq. (12.2) for the steady-state case. The functional volume contribution is

$$f = \frac{1}{2}\{ k_x(\frac{\partial T}{\partial x})^2 + k_y(\frac{\partial T}{\partial y})^2 + k_z(\frac{\partial T}{\partial z})^2\} - QT.$$

Thus, if f is to be minimized it must satisfy Eq. (12.6). Here

$$\frac{\partial f}{\partial(\partial T/\partial x)} = k_x \frac{\partial T}{\partial x}, \qquad \frac{\partial f}{\partial(\partial T/\partial y)} = k_y \frac{\partial T}{\partial y}$$

$$\frac{\partial f}{\partial(\partial T/\partial z)} = k_z \frac{\partial T}{\partial z}, \qquad \frac{\partial f}{\partial T} = -Q.$$

Equation (12.6) results in

$$\frac{\partial}{\partial x}(k_x \frac{\partial T}{\partial x}) + \frac{\partial}{\partial y} (k_y \frac{\partial T}{\partial y}) + \frac{\partial}{\partial z} (k_z \frac{\partial T}{\partial z}) + Q = 0$$

verifying that the function f does lead to correct steady state formulation, if Eq. (12.13) is also satisfied. Euler also stated that the natural boundary condition associated with Eq. (12.5) is

$$\ell_x \{\frac{\partial f}{\partial(\partial u/\partial x)}\} + \ell_y \{\frac{\partial f}{\partial(\partial u/\partial y)}\} + \ell_z \{\frac{\partial f}{\partial(\partial u/\partial z)}\} + g + hu = 0$$

on the boundary where the value of u is not forced. Here this equation simplifies to

$$\ell_x k_x \frac{\partial T}{\partial x} + \ell_y k_y \frac{\partial T}{\partial y} + \ell_z k_z \frac{\partial T}{\partial z} + g + hT = 0$$

or simply

$$k_n \frac{\partial T}{\partial n} + g + hT = 0$$

where n is the unit normal at a point on the surface and k_n is the conductivity in that direction. This is the natural boundary condition given earlier in Eq. (12.4).

12.3 Element and Boundary Matrices

From Eqs. (12.1) and (12.7) it is clearly seen that the two-dimensional functional required for the steady-state analysis is

$$I = \int_A [\frac{1}{2} \{ k_x(\frac{\partial T}{\partial x})^2 + k_y(\frac{\partial T}{\partial y})^2\} - QT]t dx\, dy + \int_\Gamma (gT + hT^2/2)t ds. \qquad (12.8)$$

where t is the thickness of the domain. We will proceed in exactly the same manner as we did for the previous variational formulations. That is, we will assume that the area integral is the sum of the integrals over the element areas. Likewise, the boundary integral where the temperature is not specified is assumed to be the sum of the boundary segment integrals. Thus

$$I = \sum_e I^e + \sum_b I^b$$

where the element contributions are

$$I^e = \int_{A^e} [\frac{1}{2}\{ k_x(\frac{\partial T}{\partial x})^2 + k_y(\frac{\partial T}{\partial y})^2\} - QT]t da$$

and the boundary segment contributions are

$$I^b = \int_{\Gamma^b} (gT + hT^2/2)t ds.$$

If we make the usual interpolation assumptions in the element and on its typical edge then we can express these quantities as

$$I^e = \frac{1}{2} \mathbf{T}^{e^T} \mathbf{S}^e \mathbf{T}^e - \mathbf{T}^{e^T} \mathbf{C}^e$$

and

$$I^b = \frac{1}{2} \mathbf{T}^{b^T} \mathbf{S}^b \mathbf{T}^b - \mathbf{T}^{b^T} \mathbf{C}^b \ .$$

Here the element matrices are

$$\mathbf{S}^e = \int_{A^e} (k_x^e \mathbf{H}_x^{e^T} \mathbf{H}_x^e + k_y^e \mathbf{H}_y^{e^T} \mathbf{H}_y^e) t^e da \tag{12.9}$$

$$\mathbf{C}_Q^e = \int_{A^e} \mathbf{H}^{e^T} Q^e t^e da \tag{12.10}$$

$$\mathbf{S}_h^b = \int_{\Gamma^b} h^b \mathbf{H}^{b^T} \mathbf{H}^b t^b ds, \quad \mathbf{C}_h^b = \int_{\Gamma^b} T_r^b h^b \mathbf{H}^{b^T} t^b ds \tag{12.11}$$

$$C_g^b = \int_{\Gamma^b} g^b \mathbf{H}^{b^T} t^b ds \tag{12.12}$$

where $\mathbf{H}$ denotes the shape functions and $\mathbf{H}_x = \partial\mathbf{H}/\partial x$, etc. For this class of problem there is only one unknown temperature per node. Once again if $\mathbf{T}$ denotes all these unknowns then $\mathbf{T}^e \subset \mathbf{T}$ and $\mathbf{T}^b \subset \mathbf{T}$.

If we select the three node (linear) triangle then the element interpolation functions, $\mathbf{H}^e$, are given in unit coordinates by Eq. (11.6) and in global coordinates by Eqs. (11.13) through (11.15). From the latter sets of equations we note that

$$\begin{aligned} \mathbf{H}_x^e &= \partial\mathbf{H}^e/\partial x = [b_1 \; b_2 \; b_3]^e/2A^e = \mathbf{d}_x \\ \mathbf{H}_y^e &= \partial\mathbf{H}^e/\partial y = [c_1 \; c_2 \; c_3]^e/2A^e = \mathbf{d}_y . \end{aligned} \tag{12.13}$$

Since these are constant we can evaluate Eq. (12.9) by inspection if the conductivities are also constant:

$$\mathbf{S}^e = \frac{k_x^e t^e}{4A^e} \begin{bmatrix} b_1 b_1 & b_1 b_2 & b_1 b_3 \\ b_2 b_1 & b_2 b_2 & b_2 b_3 \\ b_3 b_1 & b_3 b_2 & b_3 b_3 \end{bmatrix}^e \quad +$$

$$+ \frac{k_y^e t^e}{4A^e} \begin{bmatrix} c_1c_1 & c_1c_2 & c_1c_3 \\ c_2c_1 & c_2c_2 & c_2c_3 \\ c_3c_1 & c_3c_2 & c_3c_3 \end{bmatrix}^e . \qquad (12.14)$$

This is known as the element conductivity matrix. Note that this allows for different conductivities in the x- and y-directions. Equations (12.14) and (11.14) show that the conduction in the x-direction depends on the size of the element in the y-direction, and visa versa. If the internal heat generation, Q, is also constant then Eq. (12.10) can be integrated via Eq. (11.37) to yield

$$\mathbf{C}^e = \frac{Q^e A^e t^e}{3} \begin{Bmatrix} 1 \\ 1 \\ 1 \end{Bmatrix} . \qquad (12.15)$$

This internal source vector shows that a third of the internal heat, $Q^e A^e t^e$, is lumped to each of the three nodes as shown in Fig. 11.18. On a typical boundary segment the edge interpolation can be given by Eq. (5.11) or other forms. The exact integrals can be evaluated by Eq. (6.5). For example, if the coefficient, h, is constant then the boundary segment square matrix is obtained from Eq. (6.7) as

$$\mathbf{S}^b = \frac{h^b \ell^b t^b}{6} \begin{bmatrix} 2 & 1 \\ 1 & 2 \end{bmatrix} . \qquad (12.16)$$

Similarly if a constant normal flux, q, is given then the boundary flux vector is

$$\mathbf{C}^b = \frac{q^b \ell^b t^b}{2} \begin{Bmatrix} 1 \\ 1 \end{Bmatrix} . \qquad (12.17)$$

In this case half the total normal flux is lumped at each of the two nodes on the segment. This condition is analogous to that in Fig. 11.16. That figure also shows similar nodal values for higher order segments. If we select a higher order element such as the isoparametric quadrilateral then some of the above matrices are more difficult to evaluate. In the notation of Chapter 11 Eq. (12.9) becomes

$$S^e = \int_{A^e} (k_x^e \, \mathbf{d}_x^{e^T} \mathbf{d}_x^e + k_y^e \, \mathbf{d}_y^{e^T} \, \mathbf{d}_y^e) t^e da.$$

If we allow for general quadrilaterals and/or curved sides then we will need numerical integration. Thus from Eq. (11.36) we write

$$S^e = \sum_{i=1}^{NIP} W_i |J| \, (k_{x_i}^e \, \mathbf{d}_{x_i}^{e^T} \, \mathbf{d}_{x_i}^e$$

$$+ k_{y_i}^e \, \mathbf{d}_{y_i}^{e^T} \, \mathbf{d}_{y_i}^e) t_i^e \quad . \qquad (12.18)$$

Similar expressions are available for the source vectors in Eqs. (12.10) to (12.12). When the Jacobian is constant the nodal source resultants are again like those in Figs. 11.17, 11.18, 11.20, and 11.21.

12.4 Example Application

Consider a uniform square of material that has its exterior perimeter maintained at a constant temperature while its interior generates heat at a constant rate. We note that the solution will be symmetric about the square's centerlines as well as about its two diagonals. This means that we only need to utilize one-eighth of the region in the analysis. For simplicity we will assume that the material is homogeneous and $k_x = k_y = k$.

The planes of symmetry have zero normal heat flux, $q = 0$. That condition is a natural boundary condition in a finite element analysis. That is true since $\mathbf{C}_q$ in Eq. (12.12) is identically zero when the normal flux, q, is zero. The remaining essential condition is that of the known external boundary temperature as shown in Fig. 12.2. For this model we have selected four elements and six nodes. The last three nodes have the known temperature and the first three are the unknown internal temperatures. For this homogeneous region the data are:

Element	K^e	Q^e	Topology	t^e
1	8	6	1,2,3	1
2	8	6	2,4,5	1
3	8	6	5,3,2	1
4	8	6	3,5,6	1

From the geometry in Fig. 12.2 we determine the element geometric properties from Eq. (11.12). They are

	e = 1,2,4			e = 3		
i	1	2	3	1	2	3
b_i	-2	2	0	2	-2	0
c_i	0	-2	2	0	2	-2

$A^e = 2$ $\qquad$ $A^e = 2$.

From Eq. (12.14) the conduction matrix for elements 1, 2, and 4 are

$$S^e = \frac{8(1)}{4(2)} \begin{bmatrix} 4 & -4 & 0 \\ -4 & 4 & 0 \\ 0 & 0 & 0 \end{bmatrix} + \frac{8(1)}{4(2)} \begin{bmatrix} 0 & 0 & 0 \\ 0 & 4 & -4 \\ 0 & -4 & 4 \end{bmatrix}$$

or simply

$$S^e = \begin{bmatrix} 4 & -4 & 0 \\ -4 & 8 & -4 \\ 0 & -4 & 4 \end{bmatrix} . \tag{12.19}$$

Since element 3 results from a 180° rotation of element 1 it happens to have exactly the same S^e. Assembling the four element matrices gives the six system equations

$$\mathbf{ST} = \mathbf{C}$$

where

$$\mathbf{S} = \begin{bmatrix} +4 & -4 & 0 & 0 & 0 & 0 \\ -4 & (+8+4+4) & (-4-4) & -4 & 0 & 0 \\ 0 & (-4-4) & (+4+8+4) & 0 & (-4-4) & 0 \\ 0 & -4 & 0 & +8 & -4 & 0 \\ 0 & 0 & (-4-4) & -4 & (+4+4+8) & -4 \\ 0 & 0 & 0 & 0 & -4 & +4 \end{bmatrix}$$

and

$$-4T_2 + 8T_4 - 4T_5 = 4 + q_4$$

$$-4(7.75) + 8(5) - 4(5) = 4 + q_4$$

$$-15 = q_4.$$

The other two nodal fluxes are

$$q_5 = -29, \; q_6 = -4 \tag{12.23}$$

and the internal heat generated was

$$\sum_e Q^e A^e t^e = +48.$$

Thus we have verified the generated heat equals the heat outflow. Of course, this must be true for all steady state heat conduction problems.

12.5 Exercises

1. Use Fig. 11.16 to write the alternate form of Eq. (12.17) when the normal flux varies linearly from a) zero to q, b) q_1 to q_2.
2. Use Fig. 11.16 to write the alternate form of Eq. (12.17) if the uniform flux was applied to a quadratic boundary segment.
3. Use Fig. 11.18 to write the alternate form of Eq. (12.15) if we had used the six node triangle.
4. Use Fig. 11.17 to write the alternate form of Eq. (12.15) if we had used quadrilaterals with a) four nodes, b) eight nodes, c) nine nodes.
5. Verify the thermal reactions in Eq. (12.23).
6. How would the example solution in Sec. 12.4 change if we changed the topology of the third element from 5,3,2 to 3,2,5?

13. PLANE STRESS ANALYSIS

13.1 Introduction

The states of **plane stress** and **plane strain** are interesting and useful examples of stress analysis of a two-dimensional elastic solid (in the x-y plane). The assumption of plane stress implies that the component of all stresses normal to the plane are zero ($\sigma_z = \tau_{zx} = \tau_{zy} = 0$) whereas the plane strain assumption implies that the normal components of the strains are zero ($\varepsilon_z = \gamma_{zx} = \gamma_{zy} = 0$). The state of plane stress is commonly introduced in the first course of mechanics of materials. It was also the subject of some of the earliest finite element studies.

The assumption of plane stress means that the solid is very thin and that it is loaded only in the direction of its plane. At the other extreme, in the state of plane strain the dimension of the solid is very large in the z-direction. It is loaded perpendicular to the longitudinal (z) axis, and the loads do not vary in that direction. All cross-sections are assumed to be identical so any arbitrary x-y section can be used in the analysis. These two states are illustrated in Fig. 13.1. There are three common approaches to the variational formulation of the plane stress (or plane strain) problem: 1) Displacement formulation, 2) Stress formulation, 3) Mixed formulation. We will select the common displacement method and utilize the total potential energy of the system. This can be proved to be equal to assuming a Galerkin weighted residual approach. In any event note that it will be necessary to define all unknown quantities in terms of the displacements of the solid. Specifically, it will be

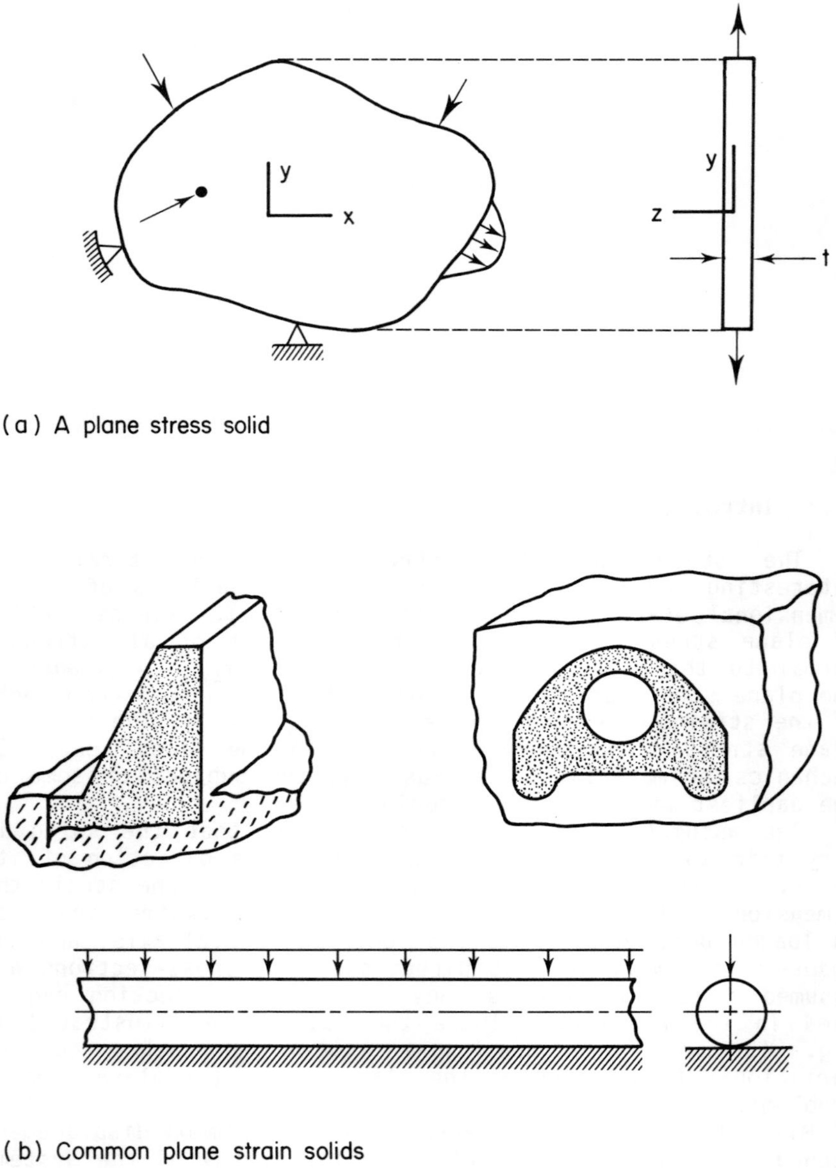

Fig. 13.1 The States of Plane Stress and Plane Strain

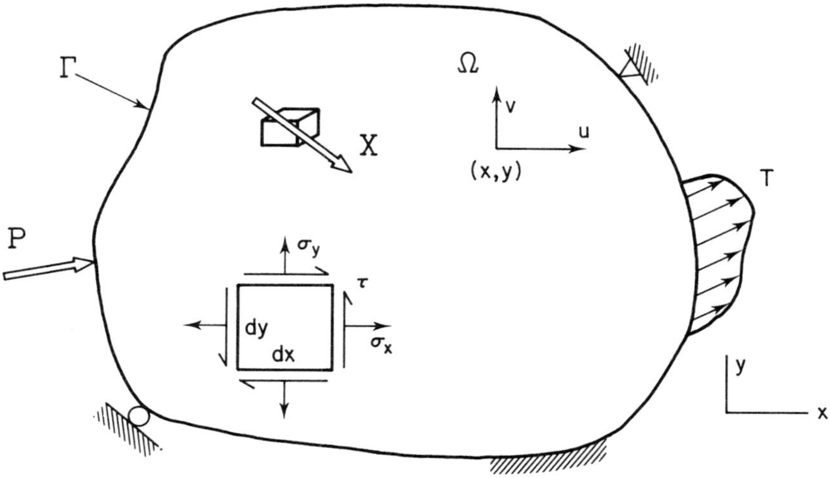

(a) The elastic solid

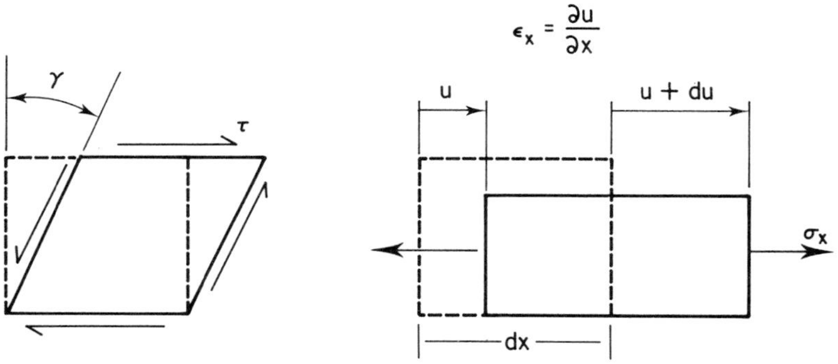

(b) Typical strains

Fig. 13.2 Notations for Plane Stress and Plane Strain

necessary to relate the strains and stresses to the displacements as was illustrated in 1-D in Sec. 4.5.

Our notation will follow that commonly used in mechanics of materials. The displacements components parallel to the x- and y-axes will be denoted by u(x,y) and v(x,y), respectively. The shear stress acting parallel to the x- and y-axes are σ_x and σ_y, respectively. The shear stress acting parallel to the y-axis on a plane normal to the x-axis is τ_{xy}, or simply τ. The corresponding components of strain are ε_x, ε_y, and γ_{xy}, or simply γ. Figure 13.2 summarizes these notations.

13.2 Minimum Total Potential Energy

Plane stress analysis, like other elastic stress analysis problems, is governed by the principle of minimizing the total potential energy in the system. These concepts were considered in Sec. 3.3 and 4.5. It is possible to write the generalized forms of the element matrices and boundary segment matrices defined in Sec. 4.5. The symbolic forms are:

(1) Stiffness matrix

$$\mathbf{S}^e = \int_{V^e} \mathbf{B}^{e^T} \mathbf{D}^e \mathbf{B}^e(x,y)dv \tag{13.1}$$

$\mathbf{B}^e$ = element strain-displacement matrix
$\mathbf{D}^e$ = material constitutive matrix

(2) Body Force Matrix

$$\mathbf{C}^e_x = \int_{V^e} \mathbf{N}^e(x,y)^T \mathbf{X}^e(x,y)dv \tag{13.2}$$

$\mathbf{N}^e$ = generalized interpolation matrix
$\mathbf{X}^e$ = body force vector per unit volume

3. Initial Strain Load Matrix

$$\mathbf{C}^e_o = \int_{V^e} \mathbf{B}^{e^T}(x,y)\ \mathbf{D}^e\ \varepsilon^e_o(x,y)dv \tag{13.3}$$

ε^e_o = initial strain matrix

4. Surface Traction Load Matrix

$$\mathbf{C}^b_T = \int_{A^b} \mathbf{N}^b(x,y)^T \, \mathbf{T}^b(x,y)da \qquad (13.4)$$

$\mathbf{N}^b$ = boundary interpolation matrix

$\mathbf{T}^b$ = traction force vector per unit surface area.

and where V^e is the element volume, A^b is a boundary segment surface area, dv is a differential volume, and da is a differential surface area. Now we will specialize these relations for plane stress (or strain). At any point the two displacement components will be denoted by $\mathbf{u}^T = [u \quad v]$. Therefore, at each node there are two displacement components (NG = 2) to be determined. The total list of element degrees of freedom will be denoted by $\mathbf{a}^e$.

13.2.1 Displacement Interpolation

As before, it is necessary to define the spatial approximation for the displacement field. Consider the x-displacement, u, at some point in an element. The simplest approximation of how it varies in space is to assume a complete linear polynomial. In two-dimensions (see Fig. 5.2) a complete linear polynomial contains three constants. Thus we select a triangular element with three nodes and assume u is to be computed at each node. Then

$$u(x,y) = \mathbf{H}^e\mathbf{u}^e$$
$$= H^e_1 \, u^e_1 + H^e_2 \, u^e_2 + H^e_3 \, u^e_3. \qquad (13.5)$$

The interpolation can either be done in global (x,y) coordinates or in a local system. If global coordinates are utilized then, from Eq. (11.13), the form of the typical interpolation function is

$$H^e_i(x,y) = (a^e_i + b^e_i x + c^e_i y)/2A^e, \; 1 \leqslant i \leqslant 3 \qquad (13.6)$$

where A^e is the area of the element and a^e_i, b^e_i, and c^e_i denote constants for node i that depend on the element geometry. Clearly we could utilize the same interpolations for the y-displacement:

$$v(x,y) = \mathbf{H}^e \, \mathbf{v}^e. \tag{13.7}$$

To define the element dof vector $\mathbf{a}^e$ we chose to order these six constants such that

$$\mathbf{a}^{e^T} = [u_1 \; v_1 \; u_2 \; v_2 \; u_3 \; v_3]^e. \tag{13.8}$$

To refer to both displacement components at a point we employ a generalized element interpolation. Then

$$\begin{Bmatrix} u(x,y) \\ v(x,y) \end{Bmatrix} = \begin{bmatrix} H_1 & 0 & H_2 & 0 & H_3 & 0 \\ 0 & H_1 & 0 & H_2 & 0 & H_3 \end{bmatrix}^e \begin{Bmatrix} u_1 \\ v_1 \\ u_2 \\ v_2 \\ u_3 \\ v_3 \end{Bmatrix}^e$$

or symbolically

$$\mathbf{u}(x,y) = \mathbf{N}^e \mathbf{a}^e \tag{13.9}$$

Of course, more advanced polynomials could be selected to define the **H** or **N** matrices.

13.2.2 Strain-Displacement Relations

From mechanics of materials we can define the strains in terms of the displacements. Order the three strain components so as to define $\varepsilon^T = [\varepsilon_x \; \varepsilon_y \; \gamma]$. These terms are defined as:

$$\varepsilon_x = \frac{\partial u}{\partial x}$$

$$\varepsilon_y = \frac{\partial v}{\partial y}$$

$$\gamma = \left(\frac{\partial u}{\partial y} + \frac{\partial v}{\partial x}\right)$$

if the common engineering form is selected for the shear strain, γ. Two of these terms are illustrated in Fig. 13.2. From Eqns. (13.5) and (13.7) we note

$$\varepsilon_x = \frac{\partial \mathbf{H}^e}{\partial x} \mathbf{u}^e$$

$$\varepsilon_y = \frac{\partial \mathbf{H}^e}{\partial y} \mathbf{v}^e$$

$$\gamma = \frac{\partial \mathbf{H}^e}{\partial y} \mathbf{u}^e + \frac{\partial \mathbf{H}^e}{\partial x} \mathbf{v}^e. \tag{13.10}$$

These can be combined into a single matrix identity to define

$$\begin{Bmatrix} \varepsilon_x \\ \varepsilon_y \\ \gamma \end{Bmatrix}^e = \begin{bmatrix} H_{1_x} & 0 & H_{2_x} & 0 & H_{3_x} & 0 \\ 0 & H_{1_y} & 0 & H_{2_y} & 0 & H_{3_y} \\ H_{1_y} & H_{1_x} & H_{2_y} & H_{2_x} & H_{3_y} & H_{3_x} \end{bmatrix}^e \begin{Bmatrix} u_1 \\ v_1 \\ u_2 \\ v_2 \\ u_3 \\ v_3 \end{Bmatrix}^e \tag{13.11}$$

or symbolically, $\boldsymbol{\varepsilon}^e = \mathbf{B}^e(x,y)\mathbf{a}^e$, where the shorthand notation $H_x = \partial H/\partial x$, etc has been employed. This defines the element strain-displacement operator $\mathbf{B}^e$ that would be used in Eqs. (13.1) and (13.3).

13.2.3 Stress-Strain Law

The stress-strain law (**constitutive relations**) between the strain components, $\boldsymbol{\varepsilon}$, and the corresponding stress components, $\boldsymbol{\sigma}^T = [\sigma_x \ \sigma_y \ \tau]$, is defined in mechanics of materials. For the case of an isotropic material in plane stress these are listed as

$$\sigma_x = \frac{E}{1 - \nu^2} (\varepsilon_x + \nu\varepsilon_y)$$

$$\sigma_y = \frac{E}{1 - \nu^2} (\varepsilon_y + \nu\varepsilon_x)$$

$$\tau = \frac{E}{2(1 + \nu)} \gamma = G\gamma \qquad (13.12)$$

where E is the elastic modulus, ν is Poisson's ratio, and G is the shear modulus. In theory, G is not an independent property. In practice it is sometimes treated as independent.

Some references list the inverse relations since the strains are usually experimentally determined from the applied stresses. In the alternate form the constitutive relations are

$$\varepsilon_x = \frac{1}{E} (\sigma_x - \nu\sigma_y)$$

$$\varepsilon_y = \frac{1}{E} (\sigma_y - \nu\sigma_x)$$

$$\gamma = \tau/G = 2\tau(1 + \nu)/E. \qquad (13.13)$$

We will write Eq. (13.12) in its matrix symbolic form

$$\boldsymbol{\sigma} = \mathbf{D}(\boldsymbol{\varepsilon} - \boldsymbol{\varepsilon}_0). \qquad (13.14)$$

Here we have allowed for the presence of initial strains, $\boldsymbol{\varepsilon}_0$, that are not usually included in mechanics of materials. For plane stress

$$\mathbf{D} = \frac{E}{1 - \nu^2} \begin{bmatrix} 1 & \nu & 0 \\ \nu & 1 & 0 \\ 0 & 0 & (1 - \nu)/2 \end{bmatrix}. \qquad (13.15)$$

Note that **D** is a symmetric matrix. This is almost always true. This observation shows that in general the element stiffness matrix, Eq. (13.1), will also be symmetric. For the sake of completeness the **constitutive matrix, D,** for plane strain will also be given. It is

$$\mathbf{D} = \frac{E(1 - \nu)}{(1 + \nu)(1 - 2\nu)} \begin{bmatrix} 1 & a & 0 \\ a & 1 & 0 \\ 0 & 0 & b \end{bmatrix}. \qquad (13.16)$$

where $a = \nu/(1 - \nu)$ and $b = (1 - 2\nu)/2(1 - \nu)$. Note that if one has an **incompressible material**, such as rubber, where $\nu = 1/2$ then division by zero would cause difficulties for plane strain problems.

The most common type of initial strain, ε_0, is that due to temperature changes. For an isotropic material these **thermal strains** are

$$\varepsilon_0 = \alpha\Delta\theta \begin{Bmatrix} 1 \\ 1 \\ 0 \end{Bmatrix} \qquad (13.17)$$

where α is the coefficient of thermal expansion and $\Delta\theta = (\theta - \theta_0)$ is the temperature rise from a stress free temperature of θ_0. Notice that thermal strains in isotropic materials do not include thermal shear strains. If the above temperature changes were present then the additional loading effects could be included via Eq. (13.3).

13.3 Matrices for the Constant Strain Triangle

Beginning with the simple linear displacement assumption of Eqs. (13.5) to (13.7) we note that for a typical interpolation function, H_i:

$$\frac{\partial H^e_i}{\partial x} = b^e_i/2A^e$$

and

$$\frac{\partial H^e_i}{\partial y} = c^e_i/2A^e.$$

Therefore, from Eqs. (13.11) and (13.12), the strain components in the triangular element are constant. Specifically

$$\mathbf{B}^e = \frac{1}{2A^e} \begin{bmatrix} b_1 & 0 & b_2 & 0 & b_3 & 0 \\ 0 & c_1 & 0 & c_2 & 0 & c_3 \\ c_1 & b_1 & c_2 & b_2 & c_3 & b_3 \end{bmatrix}^e \qquad (13.18)$$

For this reason this element is commonly known as the **constant strain triangle**, CST. If we also let the material properties, E and ν, be constant in a typical element then the stiffness matrix in Eq. (13.1) simplifies to

$$\mathbf{S}^e = \mathbf{B}^{e^T} \mathbf{D}^e \mathbf{B}^e V^e \tag{13.19}$$

where the element volume is

$$V^e = \int_{V^e} dv = \int_{A^e} t^e(x,y)dxdy \tag{13.20}$$

where t^e is the element thickness. Usually the thickness of a typical element is constant so that $V^e = t^e A^e$. Of course, it would be possible to define the thickness at each node and to utilize the interpolation functions to approximate $t^e(x,y)$.

Similarly if the temperature change in the element is also constant within the element then Eq. (13.3) defines the thermal load matrix

$$\mathbf{C}^e_o = \mathbf{B}^{e^T} \mathbf{D}^e \boldsymbol{\varepsilon}^e_o t^e A^e. \tag{13.21}$$

It would be possible to be more detailed and input the temperature at each node and integrate its change over the element.

It is common for plane stress problems to include body force loads due to gravity, centrifugal acceleration, etc. For simplicity assume that the body force vector, $\mathbf{X}^e$, and the thickness, t^e, are constant. Then the body force vector in Eq. (13.2) simplifies to

$$\mathbf{C}^e_X = t^e \int_{A^e} \mathbf{N}^{e^T} (x,y)\, dxdy\, \mathbf{X}^e. \tag{13.22}$$

From Eq. (13.9) it is noted that the non-zero terms in the integral typically involve scalar terms such as

$$\begin{aligned} I^e_i &= \int_{A^e} H^e_i(x,y)\, da \\ &= \frac{1}{2A^e} \int_{A^e} (a^e_i + b^e_i x + c^e_i y) da. \end{aligned} \tag{13.23}$$

These three terms can almost be integrated by inspection. The element geometric constants can be taken outside parts of the integrals. Then from the concepts of the first moment (centroid) of an area

$$a^e_i \int da = a^e_i A^e$$

$$\int b^e_i \, x da = b^e_i \, \bar{x}^e A^e$$

$$\int c^e_i \, y da = c^e_i \, \bar{y}^e \, A^e \qquad (13.24)$$

where $\bar{x}$ and $\bar{y}$ denote the **centroid** coordinates of the triangle. That is,

$$\bar{x}^e = (x_1 + x_2 + x_3)^e/3$$

$$\bar{y}^e = (y_1 + y_2 + y_3)^e/3.$$

In view of Eq. (13.24) the integral in Eq. (13.23) becomes

$$I^e_i = A^e(a_i + b_i\bar{x} + c_i\bar{y})^e/2A^e.$$

Reducing the algebra to its simplest form, using Eq. (11.14), yields

$$I^e_i = A^e/3, \quad 1 \leqslant i \leqslant 3. \qquad (13.25)$$

Therefore, for the CST the expanded form of Eq. (13.22) is

$$\mathbf{C}^e_X = \frac{t^e A^e}{3} \begin{bmatrix} 1 & 0 \\ 0 & 1 \\ 1 & 0 \\ 0 & 1 \\ 1 & 0 \\ 0 & 1 \end{bmatrix} \begin{Bmatrix} X_x \\ X_y \end{Bmatrix}^e = \frac{t^e A^e}{3} \begin{Bmatrix} X_x \\ X_y \\ X_x \\ X_y \\ X_x \\ X_y \end{Bmatrix}^e \qquad (13.26)$$

where X_x and X_y denote the components of the body force vector. To assign a physical meaning to this result note that $t^eA^eX^e_x$ is the resultant force in the x-direction. Therefore, the above calculation has replaced the distributed load with a statically equivalent set of three nodal loads. Each of these loads is a third of the resultant load. These **consistent loads** are illustrated in Fig. 13.3.

The final load to be considered is that acting on a typical boundary segment. As indicated in Fig. 13.3 such a segment is one side of an element being loaded with a traction. In plane stress problems these pressures or distributed shears act on the edge of the solid. In other words they are distributed over a length ℓ^b that has a known thickness, t. Those two quantities define the surface area, A^b, on which the tractions in Eq. (13.4) are applied. Similarly the differential surface area is $da = td\ell$.

We observe that such a segment would have two nodes. We can refer to them as local boundary nodes 1 and 2. Of course, they are a subset of the three element nodes and also a subset of the system nodes. Before Eq. (13.4) can be integrated to define the consistent loads on the two boundary nodes it is necessary to form the boundary interpolation, $\mathbf{N}^b$. That function defines the displacements, u and v, at all points on the boundary segment curve. By analogy with Eq. (13.8) we can denote the dof on the boundary segment as

$$\mathbf{a}^{b^T} = [u_1 \quad v_1 \quad u_2 \quad v_2]^b.$$

Then the requirement that

$$\mathbf{u} = \mathbf{N}^b\mathbf{a}^b,$$

for all points on ℓ^b, defines the required $\mathbf{N}^b$. There are actually two ways that its algebraic form can be derived:

1) Develop a consistent (linear) interpolation on the line between the nodal dof (like Eq. (4.32)), 2) Degenerate the element function $\mathbf{N}^e$, in Eq. (13.9) by restricting the x and y coordinates to points on the boundary segment.

If the second option is selected then all the H^e_i vanish except for the two associated with the two boundary segment nodes. Those two H^e_i are simplified by the restriction and thus define the two H^b_i functions. While the result of this type of procedure may be obvious the algebra is tedious in global coordinates. (For example, let $y^b = mx^b + n$ in Eq.

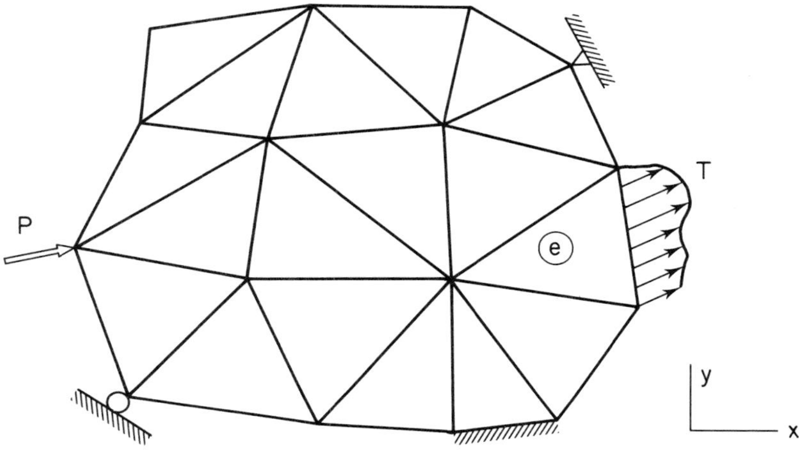

(a) Approximation of the original solid

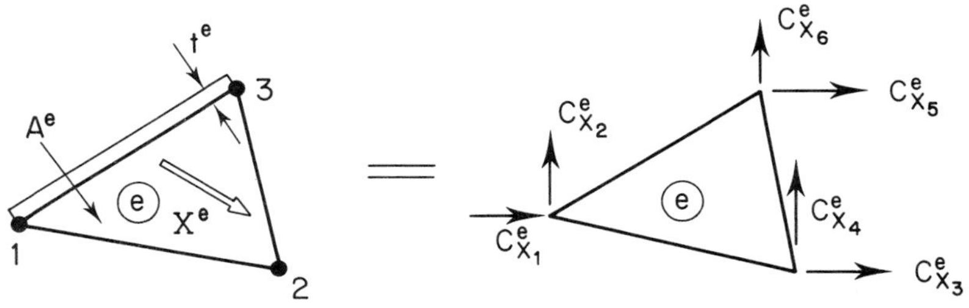

(b) Body force to element nodal load conversion

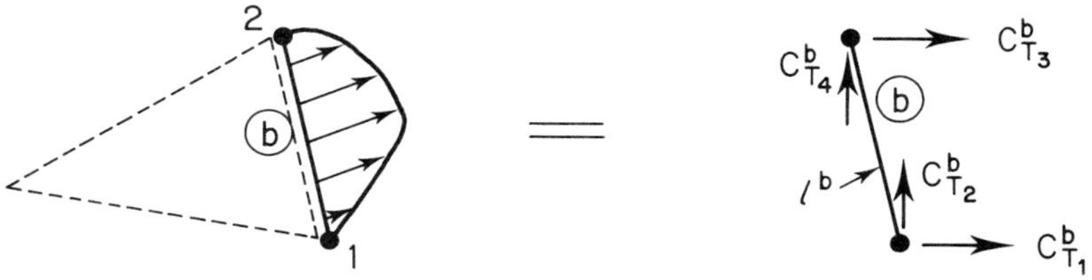

(c) Surface traction to boundary segment nodal load conversion

Fig. 13.3 Element Geometry and Consistent Loads

(13.6)). It is much easier to get the desired results if local coordinates are used. (For example, set $s^b = 0$ in Eq. (11.6)). The net result is that one obtains a one-dimensional linear interpolation set for $\mathbf{H}^b$ that is analogous to Eq. (4.18).

If we assume constant thickness, t^b, and constant tractions, $\mathbf{T}^b$, then Eq. (13.4) becomes

$$\mathbf{C}^b{}_T = t^b \int_{\ell^b} \mathbf{N}^{b^T} d\ell \, \mathbf{T}^b.$$

Repeating the procedure used for Eq. (4.32) or Eq. (4.66) a typical non-zero contribution to the integral is

$$I^b{}_i = \int_{\ell^b} H^b{}_i d\ell = \ell^b/2, \quad 1 \leqslant i \leqslant 2$$

and the final result for the four force components is

$$\mathbf{C}^b{}_T = \frac{t^b \ell^b}{2} \begin{Bmatrix} T_x \\ T_y \\ T_x \\ T_y \end{Bmatrix}^b \tag{13.27}$$

where T_x and T_y are the two coomponents of $\mathbf{T}$. Physically this states that half of the resultant x-force, $t^b \ell^b T^b{}_x$ is lumped at each of the two nodes. The resultant y-force is lumped in the same way as illustrated in Fig. 13.3.

In the case of stress analysis there are times when it is desirable to rearrange the constitutive matrix, $\mathbf{D}$. into two parts. One part, $\mathbf{D}_n$, is due to normal strain effects and the other, $\mathbf{D}_s$, is related to the shear strains. Therefore, in general it is possible to write Eq. (13.15) as

$$\mathbf{D} = \mathbf{D}_n + \mathbf{D}_s. \tag{13.28}$$

In this case such a procedure simply makes it easier to write the CST stiffness matrix in closed form. Noting that substituting Eq. (13.28) into Eq. (13.15) allows the stiffness to be separated into parts

$$\mathbf{S}^e = \mathbf{S}_n{}^e + \mathbf{S}_s{}^e$$

where

$$S_n^e = \frac{EV}{4A^2(1-\nu^2)}\begin{bmatrix} b_1^2 & & & & & \text{symmetric} \\ \nu b_1c_1 & c_1^2 & & & & \\ b_1b_2 & \nu c_1b_2 & b_2^2 & & & \\ \nu b_1c_2 & c_1c_2 & \nu b_2c_2 & c_2^2 & & \\ b_1b_3 & \nu c_1b_3 & b_2b_3 & \nu c_2b_3 & b_3^2 & \\ \nu b_1c_3 & c_1c_3 & \nu b_2c_3 & c_2c_3 & \nu b_3c_3 & c_3^2 \end{bmatrix}$$

$$S_s^e = \frac{EV}{8A^2(1+\nu)}\begin{bmatrix} c_1^2 & & & & & \text{symmetric} \\ c_1b_1 & b_1^2 & & & & \\ c_1c_2 & b_1c_2 & c_2^2 & & & \\ c_1b_2 & b_1b_2 & c_2b_2 & b_2^2 & & \\ c_1c_3 & b_1c_3 & c_2c_3 & b_2c_3 & c_3^2 & \\ c_1b_3 & b_1b_3 & c_2b_3 & b_2b_3 & c_3b_3 & b_3^2 \end{bmatrix}$$

and where V is the volume of the element. As mentioned earlier, for constant thickness $V = At$. As an example of the use of these equations consider the structure shown in Fig. 13.4. From Eq. (11.14) the element geometric constants are

e = 1			e = 2		
i	b_i	c_i	i	b_i	c_i
1	-2	-2	1	+2	+2
2	+2	0	2	-2	0
3	0	+2	3	0	-2

For the given data the constants multiplying the S_n and S_s matrices are 1×10^7 and 6×10^7, respectively. For the first element the two contributions to the element stiffness matrix are

$$S_n^e = \frac{10^7}{2}\begin{bmatrix} +8 & & & & & \text{Sym.} \\ +2 & +8 & & & & \\ -8 & -2 & +8 & & & \\ 0 & 0 & 0 & 0 & & \\ 0 & 0 & 0 & 0 & 0 & \\ -2 & -8 & +2 & 0 & 0 & +8 \end{bmatrix}$$

$$S_s^e = \frac{3 \times 10^7}{2} \begin{bmatrix} +1 & & & & & \text{Sym.} \\ +1 & +1 & & & & \\ 0 & 0 & 0 & & & \\ -1 & -1 & 0 & +1 & & \\ -1 & -1 & 0 & +1 & +1 & \\ 0 & 0 & 0 & 0 & 0 & 0 \end{bmatrix}$$

Thus the first element stiffness is:

$$\begin{matrix} & 1 & 2 & 3 & 4 & 5 & 6 & \text{local dof} \end{matrix}$$

$$S^e = 5 \times 10^6 \begin{bmatrix} +11 & & & & & \text{Sym.} \\ +5 & +11 & & & & \\ -8 & -2 & +8 & & & \\ -3 & -3 & 0 & +3 & & \\ -3 & -3 & 0 & +3 & +3 & \\ -2 & -8 & +2 & 0 & 0 & +8 \end{bmatrix}$$

$$\begin{matrix} & 1 & 2 & 5 & 6 & 3 & 4 & \text{global dof .} \end{matrix}$$

The second stiffness matrix happens to be the same due to its 180° rotation in space. Of course, its gobal dof numbers are different. That list is 7,8,3,4,5, and 6. Since there are no body forces or surface tractions these matrices can be assembled to relate the system stiffness to the applied point load, P, and the support reactions. Applying the direct assembly procedure gives

$$\begin{matrix} \text{global} & 1 & 2 & 3 & 4 & 5 & 6 & 7 & 8 \end{matrix}$$

$$5\text{X}10^6 \begin{bmatrix} +11 & & & & & & & \text{sym.} \\ +5 & +11 & & & & & & \\ -3 & -3 & +11 & & & & & \\ -2 & -8 & 0 & +11 & & & & \\ -8 & -2 & 0 & +5 & +11 & & & \\ -3 & -3 & +5 & 0 & 0 & +11 & & \\ 0 & 0 & -8 & -3 & -3 & -2 & +11 & \\ 0 & 0 & -2 & -3 & -3 & -8 & +5 & +11 \end{bmatrix} \mathbf{u} = \begin{Bmatrix} R_1 \\ R_2 \\ R_3 \\ R_4 \\ 0 \\ 0 \\ 10^4 \\ 0 \end{Bmatrix}$$

Applying the boundary conditions of zero displacement at nodes 1 and 2 reduces this set to

$$5 \text{ X } 10^6 \begin{bmatrix} 11 & & & \text{sym.} \\ 0 & 11 & & \\ -3 & -2 & 11 & \\ -3 & -8 & 5 & 11 \end{bmatrix} \begin{Bmatrix} u_3 \\ v_3 \\ u_4 \\ v_4 \end{Bmatrix} = \begin{Bmatrix} 0 \\ 0 \\ 10^4 \\ 0 \end{Bmatrix}$$

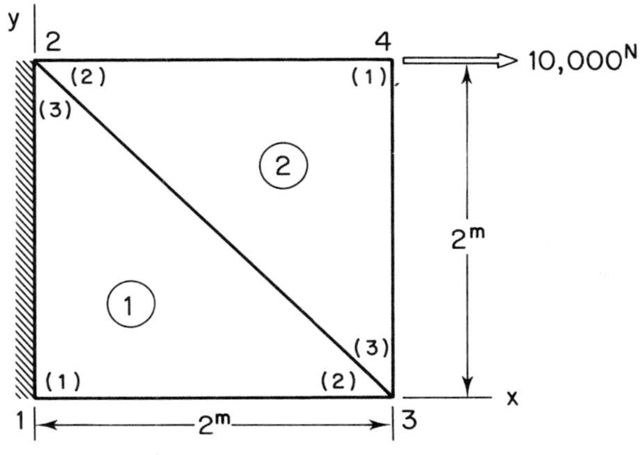

(a) Local and support conditions

element	E,(N/m^2)	t,(m)	topology			ν
1	15×10^9	5×10^{-3}	1	3	2	1/4
2	15×10^9	5×10^{-3}	4	2	3	1/4

(b) System properties

Fig. 13.4 An Example Structure

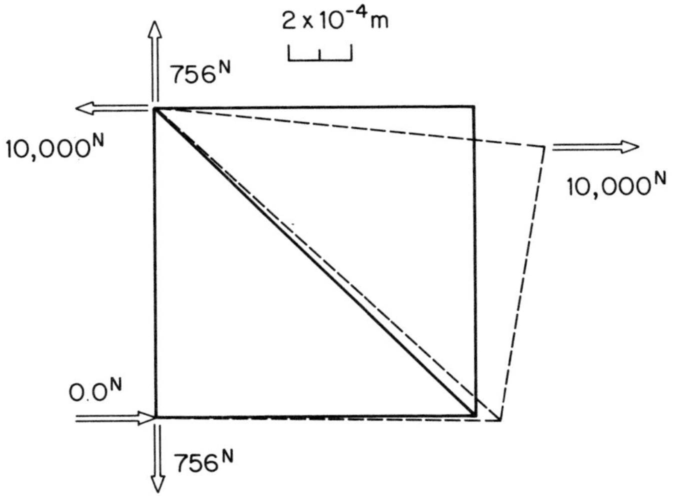

Fig. 13.5 Deformed Shape and Reactions For a Point Load

Inverting the matrix and solving gives the required displacement vector $10^5 \times \mathbf{u}_r^T$ = [2.52 -6.72 24.65 -15.41]m. Then substituting to find the reactions associated with the given displacements yields $\mathbf{R}_g^T$ = [-0.002 -756.3 -10,000 -756.3]N. The deformed shape and resulting reactions are shown in Fig. 13.5. One should always check the equilibrium of the reactions and applied loads. Checking $\Sigma \mathbf{F}_x = \mathbf{0}$, $\Sigma \mathbf{M} = \mathbf{0}$ does show minor errors in about the sixth significant figure. Thus the results are reasonable.

At this point we can recover the displacements for each element and then compute the strains and stress. The element dof vectors (in meters) are, respectively

$$\mathbf{a}^{e^T} = [0 \quad 0 \quad 2.521 \quad -6.723 \quad 0 \quad 0] \times 10^{-5}$$

$$\mathbf{a}^{e^T} = [24.650 \quad -15.406 \quad 0 \quad 0 \quad 2.521 \quad -6.723] \times 10^{-5}$$

and the strain displacement matrices, from Eq. (13.18) are

$$\mathbf{B}^e = \frac{1}{4}\begin{bmatrix} -2 & 0 & 2 & 0 & 0 & 0 \\ 0 & -2 & 0 & 0 & 0 & 2 \\ -2 & -2 & 0 & 2 & 2 & 0 \end{bmatrix}$$

for e = 1, while for e = 2

$$\mathbf{B}^e = \frac{1}{4}\begin{bmatrix} 2 & 0 & -2 & 0 & 0 & 0 \\ 0 & 2 & 0 & 0 & 0 & -2 \\ 2 & 2 & 0 & -2 & -2 & 0 \end{bmatrix} .$$

Recovering the element strains, $\boldsymbol{\varepsilon}^e = \mathbf{B}^e\,\mathbf{a}^e$ in meters/meter gives

$$e = 1, \quad \boldsymbol{\varepsilon}^{e^T} = 10^{-5}[1.261 \quad 0.0 \quad -3.361]$$

$$e = 2, \quad \boldsymbol{\varepsilon}^{e^T} = 10^{-5}[12.325 \quad -4.342 \quad 3.361]$$

Utilizing the constitute law in Eq. (13.14), with $\boldsymbol{\varepsilon}_0 = \mathbf{0}$, and

$$\mathbf{D}^e = \frac{15 \times 10^9}{(15/16)} \begin{bmatrix} 1 & 1/4 & 0 \\ 1/4 & 1 & 0 \\ 0 & 0 & 3/8 \end{bmatrix}$$

or simply

$$\mathbf{D}^e = 2 \times 10^9 \begin{bmatrix} 8 & 2 & 0 \\ 2 & 8 & 0 \\ 0 & 0 & 3 \end{bmatrix}$$

the element stresses, in Newtons/meter^2, are

$$e = 1, \quad \sigma^{e^T} = 10^4 [20.17 \quad 5.04 \quad -20.17]$$

$$e = 2, \quad \sigma^{e^T} = 10^4 [179.83 \quad -20.17 \quad 20.17] \quad .$$

A good engineer should have an estimate of the desired solution before approaching the computer. For example, if the load had been at the center of the edge then

$$\sigma_x = P/A = 10^4/(2)(5 \times 10^{-3}) = 10^6 \text{ N/m}^2,$$

and

$$\sigma_y = 0 = \tau.$$

The values are significantly different from the computed values. A better estimate would consider both the axial and bending effects so $\sigma_x = P/A \pm Mc/I$. At the centroid of these two elements ($y = 0.667$ and $y = 1.333$) the revised stress estimates are $\sigma_x = 0$ and $\sigma_x = 2 \times 10^6$ N/m^2, respectively. The revised difference between the maximum centroidal stress and our estimate is only ten percent. Of course, with the insight gained from the mechanics of materials our mesh was not a good selection. We know that while an axial stress would be constant across the depth of the member, the bending effects would vary linearly with y. Thus it was poor judgment to select a single element through the thickness.

To select a better mesh we should imagine how the stress would vary through the member. Then we would decide how many constant steps are required to get a good fit to the curve. Similarly, if we employed linear stress triangles (LST) we would estimate the required number of piece-wise linear segments needed to fit the curve. For example, consider a cantilever beam subjected to a bending load at its end. We know the exact normal stress is linear through the thickness and the shear stress varies quadratically through the thickness. Thus through the depth we would need several CST, or a few LST, or a single cubic triangle (QST).

13.4 Stress and Strain Transformations

Having computed the global stress components at a point in an element we may wish to find the stresses in another direction. This can be done by employing the transformations associated with **Mohr's circle.** Mohr's circles of stress and strain are usually used to produce graphical solutions. However, here we wish to rely on automated numerical solutions. Thus we will review the **stress transformation** laws. Refer to Fig. 13.6 where the quantities used in Mohr's transformation are defined. The alternate coordinate set (n,s) is used to describe the surfaces on which the normal stresses, σ_n and σ_s, and the shear stress, τ_{ns}, act. The n-axis is rotated from the x-axis by a positive (counter-clockwise) angle of β. By considering the equilibrium of the differential element it is shown in mechanics of materials that

$$\sigma_n = \sigma_x \operatorname{Cos}^2\beta + \sigma_y \operatorname{Sin}^2\beta + 2\tau_{xy} \operatorname{Sin}\beta \operatorname{Cos}\beta. \tag{13.28}$$

This is usually written in terms of the angle 2β so as to give

$$\sigma_n = \left(\frac{\sigma_x + \sigma_y}{2}\right) + \left(\frac{\sigma_x - \sigma_y}{2}\right) \operatorname{Cos}(2\beta) + \tau_{xy} \operatorname{Sin}(2\beta).$$

Likewise the shear stress component is found to be

$$\tau_{ns} = -\sigma_x \operatorname{Sin}\beta \operatorname{Cos}\beta + \sigma_y \operatorname{Sin}\beta \operatorname{Cos}\beta + \tau_{xy} (\operatorname{Cos}^2\beta - \operatorname{Sin}^2\beta) \tag{13.29}$$

or alternately

$$\tau_{ns} = -\left(\frac{\sigma_x - \sigma_y}{2}\right) \mathrm{Sin}^2\beta + \tau_{xy}\ \mathrm{Cos}^2\beta.$$

For Mohr's circle only these two stresses are usually plotted in the $\sigma_n - \tau_{ns}$ space. However, for a useful analytical statement we also need to define σ_s. Again from equilibrium considerations it is easy to show that

$$\sigma_s = \sigma_x\ \mathrm{Sin}^2\beta + \sigma_y\ \mathrm{Cos}^2\beta - 2\tau_{xy}\ \mathrm{Sin}\beta\ \mathrm{Cos}\beta. \tag{13.30}$$

Prior to this point we have employed matrix notation to represent the stress components. Then we were considering only the global coordinates. But now when we refer to the stress components it will be necessary to indicate which coordinate system is being utilized. We will employ the subscripts xy and ns to distinguish between the two systems. Thus our previous stress component array will be denoted by

$$\boldsymbol{\sigma}^T = \boldsymbol{\sigma}^T_{xy} = [\sigma_x\ \sigma_y\ \tau_{xy}]$$

while the new stress components will be ordered in a similar manner and denoted by

$$\boldsymbol{\sigma}^T_{ns} = [\sigma_n\ \sigma_s\ \tau_{ns}].$$

In this notation the stress transformation laws can be written as

$$\begin{Bmatrix} \sigma_n \\ \sigma_s \\ \tau_{ns} \end{Bmatrix} = \begin{bmatrix} +C^2 & +S^2 & +2SC \\ +S^2 & +C^2 & -2SC \\ -SC & +SC & (C^2-S^2) \end{bmatrix} \begin{Bmatrix} \sigma_x \\ \sigma_y \\ \tau_{xy} \end{Bmatrix} \tag{13.31}$$

where $C \equiv \mathrm{Cos}\beta$ and $S \equiv \mathrm{Sin}\beta$ for simplicity. In symbolic matrix form we can write this as

$$\boldsymbol{\sigma}_{ns} = T(\beta)\ \boldsymbol{\sigma}_{xy} \tag{13.32}$$

where T will be defined as the stress transformation matrix. Clearly if one wants to know the stresses on a given plane one specifies the angle β, forms T, and computes the results from Eq. (13.31).

A similar procedure can be employed to express Mohr's circle of strain as a strain matrix transformation law. If we denote the new strains as

$$\varepsilon_{ns}^T = [\varepsilon_n \quad \varepsilon_s \quad \gamma_{ns}]$$

then the strain transformation law is

$$\begin{Bmatrix} \varepsilon_n \\ \varepsilon_s \\ \gamma_{ns} \end{Bmatrix} = \begin{bmatrix} +C^2 & +S^2 & +SC \\ +S^2 & +C^2 & -SC \\ -2SC & +2SC & (C^2-S^2) \end{bmatrix} \begin{Bmatrix} \varepsilon_x \\ \varepsilon_y \\ \gamma_{xy} \end{Bmatrix} \tag{13.33}$$

or simply

$$\varepsilon_{ns} = \mathbf{t}(\beta)\,\varepsilon_{xy}. \tag{13.34}$$

Note that the two transformation matrices, T and **t**, are not identical. This is true because we have selected the engineering definition of the shear strain (instead of using the tensor definition). Also note that both of the transformation matrices are square. Therefore, the reverse relations can be found by inverting the transformations. That is,

$$\boldsymbol{\sigma}_{xy} = T(\beta)^{-1}\sigma_{ns}$$

and

$$\varepsilon_{xy} = \mathbf{t}(\beta)^{-1}\varepsilon_{ns}. \tag{13.35}$$

These two transformation matrices have the special property that the inverse of one is the transpose of the other. That is, it can be shown that

$$\mathbf{T}^{-1} = \mathbf{t}^T$$

and

$$\mathbf{t}^{-1} = \mathbf{T}^T. \tag{13.36}$$

This property is also true when generalized to three-dimensional properties. Another generalization is to note that if we partition the matrices into normal and shear components then

$$\mathbf{T} = \begin{bmatrix} T_{11} & T_{12} \\ T_{21} & T_{22} \end{bmatrix}$$

and

$$\mathbf{t} = \begin{bmatrix} T_{11} & T_{12}/2 \\ 2T_{21} & T_{22} \end{bmatrix} .$$

In mechanics of materials it is shown that the principle normal stresses occur when the angle is given by

$$\text{Tan } (2\beta_p) = 2\ \tau_{xy}/(\sigma_x - \sigma_y).$$

Thus if β_p were substituted into Eq. (13.31) one would compute the two principle normal stresses. In this case it may be easier to use the classical form that

$$\sigma_p = \left(\frac{\sigma_x + \sigma_y}{2}\right) \pm \left[\left(\frac{\sigma_x - \sigma_y}{2}\right)^2 + \tau_{xy}^2 \right]^{1/2}.$$

However, to illustrate the use of Eq. (13.31) we will use the results of the previous example to find the maximum normal stress at the second element centroid. Then

$$\text{Tan } (2\beta_p) = 2(20.17)/(179.83 - (-20.17)) = 0.2017$$

so $\beta_p = 5.70^0$, Cos $\beta_p = 0.995$, Sin β_p, $= 0.099$, and the transformation is

$$\begin{Bmatrix} \sigma_n \\ \sigma_s \\ \tau_{ns} \end{Bmatrix} = \begin{bmatrix} 0.9901 & 0.0099 & 0.1977 \\ 0.0099 & 0.9901 & -0.1977 \\ -0.0989 & 0.0989 & 0.9803 \end{bmatrix} \begin{Bmatrix} 179.83 \\ -20.17 \\ 20.17 \end{Bmatrix}$$

or

$$\sigma^T_{ns} = [181.84 \quad -22.18 \quad -0.00]^{N/m^2}.$$

We should also recall that the maximum shear stress is

$$\tau_{max} = (\sigma_n - \sigma_s)/2 = 102.01 \text{ N/m}^2.$$

These shear stresses occur on planes located at $(\beta_p \pm 45^0)$. The classical form for τ_{max} is

$$\tau^2_{max} = (\frac{\sigma_x - \sigma_y}{2}) + \tau^2_{xy} .$$

13.5 Anisotropic Materials

A material is defined to be **isotropic** if its material properties do not depend on direction. Otherwise it is called **anisotropic.** Most engineering materials are considered to be isotropic. However, there are many materials that are anisotropic. Examples of anisotropic materials include reinforced concrete, plywood, and filament wound fiberglass. Several special cases of anisotropic behavior have been defined. Probably the most common case is that of an **orthotropic material.** An orthotropic material has structural (or thermal) properties that can be defined in terms of two principal material axis directions. Let (n,s) be the principal material axis directions. For anisotropic materials it is usually easier to define the generalized constitutive law in the form:

$$\boldsymbol{\varepsilon}_{ns} = \mathbf{D}^{-1}_{ns}\, \boldsymbol{\sigma}_{ns} + \boldsymbol{\varepsilon}_{0\ ns} \ . \qquad (13.37)$$

Note by way comparison that Eq. (13.14) is written relative to the global coordinate axes. In Eq. (13.37) the square matrix contains the mechanical properties as experimentally measured relative to the principal material directions. For a two-dimensional orthotropic material the constitutive law is

$$\begin{Bmatrix} \varepsilon_n \\ \varepsilon_s \\ \gamma_{ns} \end{Bmatrix} = \begin{bmatrix} 1/E_n & -\nu_{sn}/E_s & 0 \\ -\nu_{ns}/E_n & 1/E_s & 0 \\ 0 & 0 & 1/G_{ns} \end{bmatrix} \begin{Bmatrix} \sigma_n \\ \sigma_s \\ \tau_{ns} \end{Bmatrix} + \boldsymbol{\varepsilon}_{0\ ns} \qquad (13.38)$$

Here the moduli of elasticity in the two principal directions are denoted by E_n and E_s. The shear modulus, G_{ns}, is independent of the elastic moduli. The two Poisson's ratios are defined by the following notation:

$$\nu_{ij} = \varepsilon_i / \varepsilon_j \qquad (13.39)$$

where i denotes the direction of the load, ε_i is the normal strain in the load direction; and ε_j is the normal strain in the transverse (orthogonal) direction. Symmetry considerations result in the additional requirement that

$$E_n \nu_{sn} = E_s \nu_{ns}. \qquad (13.40)$$

Thus four independent constants must be measured to define the orthotropic material mechanical properties. If the material is isotropic then

$$\nu = \nu_{ns} = \nu_{sn},$$

$$E = E_s = E_n,$$

and

$$G = G_{ns} = E/(2(1 + \nu)).$$

In that case only two constants (E and ν) are required and they can be measured in any direction. When the material is isotropic then Eq. (13.38) reduces to Eq. (13.13).

Orthotropic materials also have thermal properties that vary with direction. If $\Delta\theta$ denotes the temperature change then the local initial thermal strain is

$$\boldsymbol{\varepsilon}_{0\,ns}^{T} = \Delta\theta[\alpha_n \;\; \alpha_s \;\; 0]$$

where α_n and α_s are the principal coefficients of thermal expansion. If one is given the orthotropic properties it is common to numerically invert $\mathbf{D}_{ns}^{-1}$ to give the usual form

$$\boldsymbol{\sigma}_{ns} = \mathbf{D}_{ns} (\boldsymbol{\varepsilon}_{ns} - \boldsymbol{\varepsilon}_{0\,ns}). \tag{13.41}$$

This is done since this algebraic form of $\mathbf{D}_{ns}$ is algebraically much more complicated than its inverse. Due to experimental error in measuring the anisotropic constants there is a potential difficulty with this concept. For a physically possible material it can be shown that both must be positive definite. This means that the determinant must be greater than zero. Due to experimental error it is not unusual for this condition to be violated. When this occurs the program should be designed to stop and require acceptable data. Then the user must adjust the experimental data to satisfy the condition that

$$(E_n - E_s \nu_{ns}^2) > 0.$$

From the previous section on stress and strain transformations we know how to obtain $\boldsymbol{\varepsilon}_{xy}$ and $\boldsymbol{\sigma}_{xy}$ from given $\boldsymbol{\varepsilon}_{ns}$ and $\boldsymbol{\sigma}_{ns}$. But how do we obtain $\mathbf{D}_{xy}$ from $\mathbf{D}_{ns}$? We must have $\mathbf{D}_{xy}$ to form the stiffness matrix since it must be integrated relative to the x - y axes. That is,

$$\mathbf{K}_{xy} = \int \mathbf{B}_{xy}^{T} \, \mathbf{D}_{xy} \, \mathbf{B}_{xy} dv.$$

Thus the use of the n - s axes, in Fig. 13.6, to define (input) the material properties requires that we define one more transformation law. It is the transformation from $\mathbf{D}_{ns}$ to $\mathbf{D}_{xy}$. There are various ways to derive the required transformation. One simple procedure is to recall that the

$$\mathbf{D}_{xy} = \mathbf{t}_{ns}^{T} D_{ns}\ \mathbf{t}_{ns}. \tag{13.44}$$

The same concept holds for general three-dimensional problems.

Before leaving the concept of anisotropic materials we should review the initial thermal strains. Recall that for an isotropic material or for an anisotropic material in principal axes a change in temperature does not induce an initial shear strain. However, an anisotropic material does have an initial thermal shear strain in other coordinate directions. From Eqs. (13.35) and (13.36) we have

$$\boldsymbol{\varepsilon}_{0\ xy} = \mathbf{t}_{ns}\ \boldsymbol{\varepsilon}_{0\ ns}$$

so for an orthotropic material

$$\begin{Bmatrix} \varepsilon_x^0 \\ \varepsilon_y^0 \\ \gamma_{xy}^0 \end{Bmatrix} = \begin{bmatrix} +C^2 & +S^2 & -SC \\ +S^2 & +C^2 & +SC \\ +2SC & -2SC & (C^2-S^2) \end{bmatrix} \begin{Bmatrix} \varepsilon_n^0 \\ \varepsilon_s^0 \\ 0 \end{Bmatrix}.$$

Thus the shear strain is

$$\gamma_{xy}^0 = 2\ \mathrm{Sin}\ \beta\ \mathrm{Cos}\ \beta(\varepsilon_n^0 - \varepsilon_s^0).$$

This is not zero unless the two axes systems are the same ($\beta = 0$ or $\beta = \pi/2$). Therefore, one must replace the previous null terms in Eqs. (13.17) and (13.21).

13.6 Exercises

1. Verify the $\mathbf{t}^{-1} = T^T$ identity by computing the product $\mathbf{tT}^T$ and comparing with the identity matrix.
2. Find the strains in the CST when it is given a rigid body displacement in the x-direction. That is, when $\mathbf{u}^T = [c\ 0\ c\ 0\ c\ 0]$ where c is an arbitrary constant.
3. Repeat the stress analysis example in Sec. 13.3 when the load vector results from a constant edge traction of $T_x = 1000\ N/m^2$ on edge 3-4.

strain energy density is a scalar. Therefore, it must be the same in all coordinate systems. The strain energy density at a point is

$$U = 1/2\ \boldsymbol{\sigma}^T\ \boldsymbol{\varepsilon} = 1/2\ \boldsymbol{\varepsilon}^T \boldsymbol{\sigma}.$$

In the global axes it is

$$U = 1/2\ \boldsymbol{\sigma}_{xy}^{\ T}\ \boldsymbol{\varepsilon}_{xy} = 1/2\ (\mathbf{D}_{xy}\,\boldsymbol{\varepsilon}_{xy})^T\,\boldsymbol{\varepsilon}_{xy}^{\ T}$$

$$= 1/2\ \boldsymbol{\varepsilon}_{xy}^T\ \mathbf{D}_{xy}\ \boldsymbol{\varepsilon}_{xy}. \qquad (13.42)$$

In the principal material directions it is

$$U = 1/2\ \boldsymbol{\sigma}_{ns}^T\ \boldsymbol{\varepsilon}_{ns} = 1/2\ (\mathbf{D}_{ns}\ \boldsymbol{\varepsilon}_{ns})^T\,\boldsymbol{\varepsilon}_{ns}$$

$$= 1/2\ \boldsymbol{\varepsilon}_{ns}^T\,\mathbf{D}_{ns}\ \boldsymbol{\varepsilon}_{ns}.$$

But from our Mohr's circle transformation for strain (Eq. (13.34))

$$\boldsymbol{\varepsilon}_{ns} = \mathbf{t}_{ns}\ \boldsymbol{\varepsilon}_{xy}$$

so in the n - s axes

$$U = 1/2\ (\mathbf{t}_{ns}\ \boldsymbol{\varepsilon}_{xy})^T\ \mathbf{D}_{ns}\ (\mathbf{t}_{ns}\ \boldsymbol{\varepsilon}_{xy})$$

$$= 1/2\ \boldsymbol{\varepsilon}_{xy}^T(\mathbf{t}_{ns}^T\mathbf{D}_{ns}\ \mathbf{t}_{ns})\ \boldsymbol{\varepsilon}_{xy}. \qquad (13.43)$$

Comparing Eqs. (13.42) and (13.43) gives the **constitutive transformation** law that

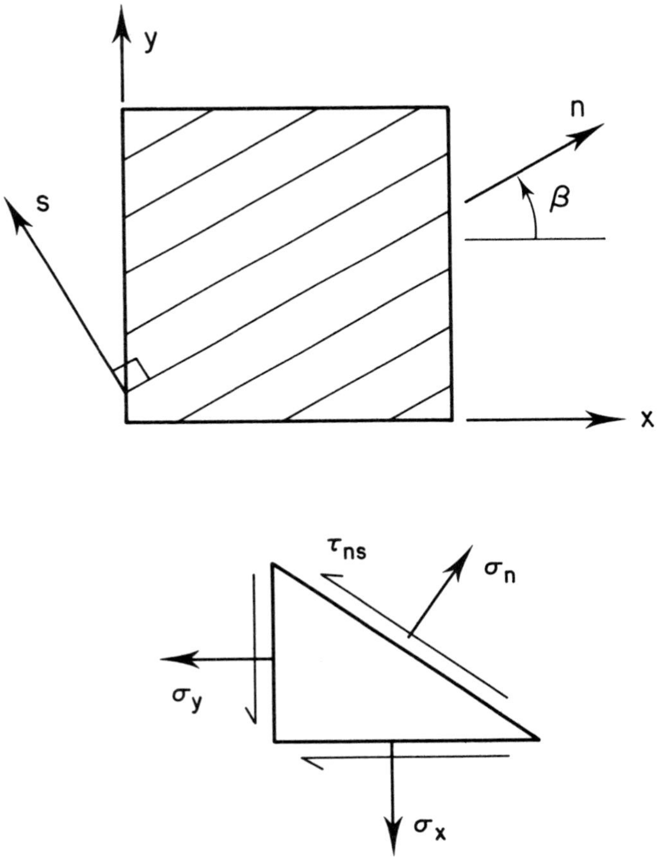

Fig. 13.6 Local Material or Stress Axes

14. AXISYMMETRIC ANALYSIS

14.1 Introduction

There are many problems that can accurately be modeled as being revolved about an axis. Many of these can be analyzed by employing a radial coordinate, R, and an axial coordinate, Z. Solids of revolution can be formulated in terms of the two-dimensional area that is revolved about the axis. Numerous other objects are very long in the axial direction and can be treated as segments of a cylinder. This reduces the analysis to a one-dimensional study in the radial direction. We will begin with that common special case.

We will find that changing to these **cylindrical coordinates** will make small changes in the governing differential equations and the corresponding integral theorems that govern the finite element formulation. Also the volume and surface integrals take on special forms. These use the **Theorems of Pappus.** The first states that the surface area of a revolved arc is the product of the arc length and the distance traveled by the centroid of the arc. The second states that the volume of revolution of the generating area is the product of the area and the distance traveled by its centroid. In both cases the distance traveled by the centroid, in full revolution, is $2\pi\bar{R}$ where $\bar{R}$ is the centroidal radial coordinate of the arc or area. If we consider differential arcs or areas then the corresponding differential surface or volume of revolution are

$$da = 2\,\pi\,R\,d\ell \qquad (14.1)$$

$$dv = 2\,\pi\,R\,dR\,dZ. \tag{14.2}$$

14.2 Heat Conduction in a Cylinder

The previous one-dimensional heat transfer model becomes more complicated here. When we consider a point on a radial line we must remember that it is a cross-section of a ring of material around the hoop of the cylinder. Thus as heat is conducted outward in the radial direction it passes through an ever increasing amount of material. The resulting differential equation for thermal equilibrium is well known:

$$\frac{1}{R}\frac{d}{dR}\left(Rk\,\frac{dT}{dR}\right) + Q = 0 \tag{14.3}$$

where R is the radial distance from the axis of revolution, k is the thermal conductivity, T is the temperature, and Q is the internal heat generation per unit volume. One can have essential boundary conditions where T is given or as a surface flux condition

$$-k\,\frac{dT}{dR} = q \tag{14.4}$$

where q is the flux normal to the surface (i.e. radially). If we multiply Eq. (14.3) by R it would look like our previous one-dimensional form:

$$\frac{d}{dR}\left(k^{*}\,\frac{dT}{dR}\right) + Q^{*} = 0$$

where

$$k^{*} = Rk$$

$$Q^{*} = RQ$$

could be viewed as variable coefficients. This lets us find the required integral (variational) form by inspection. It is

$$I = 2\pi\Delta Z\,\int\left[\frac{1}{2}\,k^{*}\left(\frac{dT}{dR}\right)^{2} - Q^{*}T\right]dR \qquad \rightarrow \text{min}$$

or

$$I = 2\pi\Delta Z\,\int\frac{1}{2}\,\left[k\left(\frac{dT}{dR}\right)^{2} - QT\right]RdR \qquad \rightarrow \text{min} \tag{14.5}$$

where the integration limits are the inner and outer radii of the cylindrical segment under study. The typical length in the axial direction, ΔZ, is usually defaulted to unity. The corresponding element conduction matrix is

$$\mathbf{S}^e = 2\pi \int_{\ell^e} k^e \frac{d\mathbf{H}^{e^T}}{dR} \frac{d\mathbf{H}^e}{dR} RdR \tag{14.6}$$

and the source vector (if any) is

$$\mathbf{C}_Q^e = 2\pi \int_{\ell^e} \mathbf{H}^{e^T} Q^e RdR. \tag{14.7}$$

If we consider a two node (linear) line element in the radial direction we can use our previous results to write these matrices by inspection. Recalling that $\ell^e = (R_2 - R_1)^e$ and

$$\frac{d\mathbf{H}^e}{dR} = \frac{1}{\ell^e}[-1 \quad 1]$$

and assuming a constant material property in the element gives

$$\mathbf{S}^e = 2\pi \frac{k^e}{(\ell^e)^2} \begin{bmatrix} 1 & -1 \\ -1 & 1 \end{bmatrix} \int_{R_1}^{R_2} RdR$$

$$\mathbf{S}^e = 2\pi \frac{k^e (R_2{}^2 - R_1{}^2)^e}{2(\ell^e)^2} \begin{bmatrix} 1 & -1 \\ -1 & 1 \end{bmatrix} \tag{14.8}$$

or, in an alternate form

$$\mathbf{S}^e = \pi \frac{k^e (R_2 + R_1)^e}{(R_2 - R_1)^e} \begin{bmatrix} 1 & -1 \\ -1 & 1 \end{bmatrix} \tag{14.9}$$

Thus, unlike the original one-dimensional case the conduction matrix depends on where the element is located. That is, it depends on how much material it includes (per unit length in the axial direction). Next we assume a constant source term so

$$\mathbf{C}_Q^e = 2\pi Q^e \int_{\ell^e} \mathbf{H}^{e^T} RdR.$$

But the interpolation functions, **H**, also depend on the radial position. One approach to this integration is to again use our isoparametric interpolation and let

$$R = \mathbf{H}^e \mathbf{R}^e.$$

Then

$$\mathbf{C}_Q^e = 2\pi Q^e \left(\int_{\ell^e} \mathbf{H}^{e^T} \mathbf{H}^e dR \right) \mathbf{R}^e \tag{14.10}$$

Therefore,

$$\mathbf{C}_Q^e = 2\pi \frac{Q^e \ell^e}{6} \begin{bmatrix} 2 & 1 \\ 1 & 2 \end{bmatrix} \begin{Bmatrix} R_1 \\ R_2 \end{Bmatrix}^e$$

$$= 2\pi \frac{Q^e \ell^e}{6} \begin{bmatrix} 2R_1 + R_2 \\ R_1 + 2R_2 \end{bmatrix}^e . \tag{14.11}$$

As a simple numerical example consider a cylinder with constant properties, no internal heat generation, an inner radius temperature of $T = 100$ at $R = 1$ and an outer radius temperature of $T = 10$ at $R = 2$. Select a model with four equal length elements and five nodes. Numbering the nodes radially we have essential boundary conditions of $T_1 = 100$ and $T_5 = 10$. Considering the form in Eq. (14.9) we note that the element values of $(R_2 - R_1)^e/\ell^e$ are 9, 11, 13, and 15, respectively. Therefore we can write the assembled system equations as

$$\pi k^e \begin{bmatrix} 9 & -9 & 0 & 0 & 0 \\ -9 & (9+11) & -11 & 0 & 0 \\ 0 & -11 & (11+13) & -13 & 0 \\ 0 & 0 & -13 & (13+15) & -15 \\ 0 & 0 & 0 & -15 & 15 \end{bmatrix} \begin{Bmatrix} T_1 \\ T_2 \\ T_3 \\ T_4 \\ T_5 \end{Bmatrix} = \begin{Bmatrix} q_1 \\ 0 \\ 0 \\ 0 \\ -q_5 \end{Bmatrix} .$$

Applying the essential boundary conditions, and dividing both sides by the leading constant gives the reduced set

$$\begin{bmatrix} 20 & -11 & 0 \\ -11 & 24 & -13 \\ 0 & -13 & 28 \end{bmatrix} \begin{Bmatrix} T_2 \\ T_3 \\ T_4 \end{Bmatrix} = 100 \begin{Bmatrix} 9 \\ 0 \\ 0 \end{Bmatrix} + 10 \begin{Bmatrix} 0 \\ 0 \\ 15 \end{Bmatrix} .$$

Solving yields the internal temperature distribution of $T_2 = 71.06$, $T_3 = 47.39$, and $T_4 = 27.34$. Comparing with the exact solution

$$T = [T_1 \ln(R_5/R) + T_5 \ln(R/R_1)]/\ln(R_5/R_1)$$

shows that our approximation is accurate to at least three significant figures. Also note that both the exact and approximate temperature distributions are independent of the thermal conductivity, k. This is true only because the internal heat generation, Q, was zero. Of course, k does have some effect on the external heat fluxes (thermal reactions), q_1, and q_5, necessary to maintain the prescribed surface temperatures. Substituting back into the first equation to recover the thermal reaction we obtain

$$[9(100) - 9(71.06) + 0] = q_1/\Pi k^e$$

$$818.3 = q_1/\Pi k^e$$

entering at the inner radius. The fifth equation gives q_5 an equal amount exiting at the outer radius. Therefore, in this problem the heat flux is always in the positive radial direction. It should be noted that if we had used a higher order element then the integrals would have been much more complicated than the one-dimensional case. This is typical of most axisymmetric problems. Of course, in practice we use numerical integration to automate the evaluation of the element matrices.

14.3 General Field Problems

If the axial direction is also important in the analysis then Eq. (14.3) generalizes to

$$\frac{1}{R}\frac{\partial}{\partial R}(Rk_R \frac{\partial T}{\partial R}) + k_Z \frac{\partial^2 T}{\partial Z^2} + Q = 0. \tag{14.12}$$

The corresponding element conduction matrix is

$$\mathbf{S}^e = 2\pi \int_{A^e} (k_R \frac{\partial \mathbf{H}^T}{\partial R} \frac{\partial \mathbf{H}}{\partial R} + k_Z \frac{\partial \mathbf{H}^T}{\partial Z} \frac{\partial \mathbf{H}}{\partial Z}) RdRdZ \tag{14.13}$$

and the source vector is

$$\mathbf{C}_Q^e = 2\pi \int_{A^e} \mathbf{H}^T Q^e RdRdZ. \tag{14.14}$$

If we again consider the simple three node triangle and recall that it has constant global derivatives we can again integrate Eq. (14.13) by inspection. The constant matrix is removed from the integral and we are left only with the volume calculation from the Theorem of Papus:

$$V^e = 2\pi \int_{A^e} R dR dZ = 2\pi \int_{A^e} R dA$$

$$= 2\pi \bar{R} A^e$$

where $\bar{R} = (R_1 + R_2 + R_3)^e/3$ is the radial centroid of the element. Before, in Eq. (12.14), the planar element had a volume of $V = tA$ where t was the thickness. Here we simply have to replace the previous t with $2\pi\bar{R}$.

Even with a constant source Eq. (14.14) is not trivial. If we interpolate the radial position from the nodal data:

$$R = \mathbf{H}\mathbf{R}^e$$

then

$$\mathbf{C}_Q^e = 2\pi Q^e \int_{A^e} \mathbf{H}^T\mathbf{H}\, dRdZ\ \mathbf{R}^e.$$

The integral was previously evaluated for this element. Substituting the matrix integral gives

$$\mathbf{C}_Q^e = \frac{2\pi Q^e A^e}{12}\begin{bmatrix} 2 & 1 & 1 \\ 1 & 2 & 1 \\ 1 & 1 & 2 \end{bmatrix}\begin{Bmatrix} R_1 \\ R_2 \\ R_3 \end{Bmatrix}^e . \tag{14.15}$$

14.4 Axisymmetric Stress Analysis

Another common problem is the analysis of an axisymmetric solid with axisymmetric loads and supports. This becomes a two-dimensional analysis that is very similar to the plane strain analysis considered earlier. The radial and axial displacement components will be denoted by u and v. These are the same unknowns used in the plane strain study.

In addition to the previous three strains there is a fourth strain known as the **hoop strain**, ε_θ. There is a corresponding hoop stress, σ_θ.

The hoop strain results from the change in length of a fiber of material around the circumference of the solid. Recall the definition of strain as a change in length divided by the original length. The circumference at a typical radial position is $L = 2\pi R$. When such a point undergoes a radial displacement of u it occupies a new radial position of (R +

u). It has a corresponding increase in circumference. The hoop strain becomes

$$\varepsilon_\theta = \frac{\Delta L}{L} = \frac{2\pi(R + u) - 2\pi R}{2\pi R}$$

$$\varepsilon_\theta = u/R. \tag{14.16}$$

Then our strains are denoted as

$$\boldsymbol{\varepsilon}^T = [\varepsilon_R \ \varepsilon_Z \ \varepsilon_\theta \ \gamma] \tag{14.17}$$

and the corresponding stress components are

$$\boldsymbol{\sigma}^T = [\sigma_R \ \sigma_Z \ \sigma_\theta \ \tau]. \tag{14.18}$$

For an isotropic material the stress-strain law is like that for the plane strain case. These relations are

$$\varepsilon_R = \frac{1}{E}(\sigma_R - \nu\sigma_Z - \nu\sigma_\theta)$$

$$\varepsilon_Z = \frac{1}{E}(\sigma_Z - \nu\sigma_R - \nu\sigma_\theta)$$

$$\varepsilon_\theta = \frac{1}{E}(\sigma_\theta - \nu\sigma_R - \nu\sigma_Z) \tag{14.19}$$

$$\gamma = \frac{2(1 + \nu)}{E} \tau.$$

Solving for the three normal stresses in terms of the three normal strains and expressing in matrix form gives

$$\boldsymbol{\sigma} = \mathbf{D} \ \boldsymbol{\varepsilon}$$

where

$$D = \frac{E}{(1 + \nu)(1 - 2\nu)} \begin{bmatrix} a & \nu & \nu & 0 \\ \nu & a & \nu & 0 \\ \nu & \nu & a & 0 \\ 0 & 0 & 0 & b \end{bmatrix}$$

with

$$a = (1 - \nu)$$

$$b = (1 - 2\nu)/2.$$

The strain-displacement matrix is like the plane strain problem plus an additional row for the hoop effects. Writing u in terms of the interpolation function yields that new row. That is,

$$\varepsilon_\theta(R,Z) = u(R,Z)/R$$
$$= \mathbf{H}^e \mathbf{u}^e / R.$$

Using our previous notation for the strain-displacement relation we have

$$\boldsymbol{\varepsilon}^e = \mathbf{B}^e \mathbf{a}^e.$$

Here a typical nodal contribution to B^e is

$$\mathbf{B}_i^e = \begin{bmatrix} \partial H_i / \partial R & 0 \\ 0 & \partial H_i / \partial Z \\ H_i / R & 0 \\ \partial H_i / \partial Z & \partial H_i / \partial R \end{bmatrix}$$

where i is the local node number. Since H is now present in B it should be clear that the integration of B to form **K** will be more difficult than the plane strain case. Even if we use the three node triangular element only three of the four strains are constant. The hoop contribution always varies with the radius. The stiffness matrix is given by the volume integral

$$\mathbf{K}^e = 2\pi \int_{A^e} \mathbf{B}^{e^T} \mathbf{D}^e \mathbf{B}^e R dR dZ \quad .$$

The terms in this equation can be integrated in closed form for triangular elements. However, these equations are complicated and it is simpler to program using numerical integration. For a small element it is reasonable to approximate the integral by using a constant B matrix based on its value at the centroid of the element. Then

$$\mathbf{K}^e \approx 2\pi\, \overline{\mathbf{B}}^e \; \mathbf{D}^e \; \overline{\mathbf{B}}^e \int_{A^e} R dR dZ$$

$$= 2\pi \overline{R}\; A^e \; \overline{\mathbf{B}}^e \; \mathbf{D}^e \; \overline{\mathbf{B}}^e$$

where a bar denotes a quantity evaluated at the centroid. This can also be viewed as a one point numerical integration rule. For curved elements, or elements with large changes in the radius it is necessary to employ a more accurate integration rule.

14.5 Structural Analysis of a Cylinder

In the case of an infinite cylinder the above formulation simplifies since v = 0 and $\partial/\partial z = 0$. Thus we consider only the radial displacement, u, and the strains and stresses in the radial and hoop directions. This gives the two strains as

$$\boldsymbol{\varepsilon} = \mathbf{B}^e \mathbf{u}^e$$

where

$$\mathbf{B}^e = \begin{bmatrix} \partial \mathbf{H}/\partial R \\ \mathbf{H}/R \end{bmatrix} . \tag{14.21}$$

The radial and hoop stresses are

$$\boldsymbol{\sigma} = \mathrm{D}\ \boldsymbol{\varepsilon}$$

where

$$\mathbf{D} = \begin{bmatrix} D_{11} & D_{12} \\ D_{12} & D_{22} \end{bmatrix} \tag{14.22}$$

and for an isotropic material

$$D_{11} = D_{22} = E(1 - \nu)/(1 + \nu)(1 - 2\nu)$$

$$D_{12} = D_{21} = E\nu/(1 + \nu)(1 - 2\nu)$$

The stiffness matrix in Eq. (14.20) can be expanded to the form

$$\mathbf{K}^e = 2\pi\ \Delta z \int_{\ell^e} [D_{11} \frac{\partial \mathbf{H}^T}{\partial R} \frac{\partial \mathbf{H}}{\partial R}$$

$$+ D_{12}(\frac{\partial \mathbf{H}^T}{\partial R} \mathbf{H} + \mathbf{H}^T \frac{\partial \mathbf{H}}{\partial R})/R \tag{14.23}$$

$$+ D_{22}\ \mathbf{H}^T\ \mathbf{H}/R^2]RdR.$$

The first integral we just evaluated and is given in Eqs. (14.8) and (14.9) if we let $\Delta z = 1$ and replace k with D_{11}. The second term we integrate by inspection since the R terms cancel. The result is

$$\mathbf{K}_{12}^{e} = 2\pi\ \Delta z\ D_{12}^{e}\begin{bmatrix} -1 & 0 \\ 0 & 1 \end{bmatrix} . \tag{14.24}$$

The remaining contribution is more difficult since it involves division by R. Assuming constant D_{22} we have

$$\mathbf{K}_{22} = 2\pi\ \Delta z\ D_{22} \int_{\ell^e} \frac{1}{R}\mathbf{H}^T\mathbf{H}\ dR \tag{14.25}$$

which requires analytic integration involving logarithms, or numerical integration. Using a one point (centroidal) rule gives

$$\mathbf{K}_{22}^{e} = \frac{2\pi\ \Delta z\ D_{22}^{e}\ell^{e}}{2(R_1 + R_2)^{e}}\begin{bmatrix} 1 & 1 \\ 1 & 1 \end{bmatrix} . \tag{14.26}$$

For a cylinder the loading would usually be a pressure acting on an outer surface or an internal centrifugal load. For a pressure load the resultant force at a nodal ring is the pressure times the surface area. Thus

$$F_{p_i} = 2\pi\ \Delta z\ R_i p_i\ .$$

As a numerical example consider a single element solution for a cylinder with an internal pressure of p = 1 ksi on the inner radius R_1 = 10 in. Assume $E = 10^4$ ksi and $\nu = 0.3$, and let the thickness of the cylinder by 1 in. Note that there is no essential boundary condition on the radial displacement. This is because the hoop effects prevent a rigid body radial motion. The numerical values of the above stiffness contributions are

$$\mathbf{K}_{11} = 2\pi\ \Delta z(1.413 \times 10^5)\begin{bmatrix} 1 & -1 \\ -1 & 1 \end{bmatrix}$$

$$\mathbf{K}_{12} = 2\pi\ \Delta z(5.769 \times 10^3)\begin{bmatrix} -1 & 0 \\ 0 & 1 \end{bmatrix}$$

$$\mathbf{K}_{22} = 2\pi\ \Delta z(3.205 \times 10^2)\begin{bmatrix} 1 & 1 \\ 1 & 1 \end{bmatrix}$$

while the resultant force at the inner radius is

$$F_p = 2\pi\ \Delta z\ 10.$$

Assembling and canceling the common constant gives

$$10^5 \begin{bmatrix} 1.35897 & -1.41026 \\ -1.41026 & 1.47436 \end{bmatrix} \begin{Bmatrix} u_1 \\ u_2 \end{Bmatrix} = \begin{Bmatrix} 10 \\ 0 \end{Bmatrix}$$

Solving gives $\mathbf{u} = [9.9642 \quad 9.5309] \times 10^{-3}$ in. This represents a displacement error of about 8% and 9%, respectively at the two nodes. The maximum radial stress equals the applied pressure. The stresses can be found from Eq. (14.22). The hoop strain at node 1 is

$$\varepsilon_\theta = u_1/R_1 = 0.064 \times 10^{-4} \text{ in/in.}$$

The radial strain is

$$\varepsilon_R = \frac{\partial H_1}{\partial R} u_1 + \frac{\partial H_2}{\partial R} u_2.$$

The constant radial strain approximation is

$$\varepsilon_R = \frac{-u_1 + u_2}{R_2 - R_1} = -4.333 \times 10^{-4} \text{ in/in.}$$

Therefore the hoop stress at the first node is

$$\sigma_\theta = D_{12}\varepsilon_R + D_{22}\varepsilon_\theta$$

$$\sigma_\theta = -2.500 + 13.413$$

$$\sigma_\theta = 10.91 \text{ ksi.}$$

This compares well with the exact value of 10.52 ksi. Note that the inner hoop stress is more than ten times the applied internal pressure.

14.6 Exercises

1. Integrate Eq. (14.25) exactly for linear interpolation where $\mathbf{H} = [(R_2 - R) \quad (R - R_1)]/\ell$.
2. Repeat the cylinder stress example using a) two elements using Eq. (14.26), b) one element using the exact form of Eq. (14.25).

15 SIMPLE HARMONIC MOTION AND EIGENVALUE PROBLEMS

15.1 Introduction

Another common class of analysis is the solution of eigenvalue problems. These occur in many engineering applications. Typical examples are structural vibrations, tidal oscillations, acoustical pressure distributions, electromagnetic waves, etc. The typical problem is described by one or more characteristic values or **eigenvalues**, and a corresponding spatial distributions or mode vectors or **eigenvectors**. One of the most common examples is the simple harmonic motion of an elastic system. Thus we will begin with that case.

15.2 The Mass Matrix

Consider a simple spring-mass system subjected to an external force as shown in Fig. 15.1. We have previously treated elastic stiffnesses and external forces, but this is the first time we have treated the effect of a mass. Those earlier treatments introduced the concepts of strain energy (potential energy) and work by external forces. From our introductory courses in particle dynamics we should recall that a moving mass has an additional type of energy associated with it: **kinetic energy.** We defined the kinetic energy of a particle as

$$T = \frac{1}{2} mv^2 \tag{15.1}$$

where v is the speed of the particle, i.e. $v = \dot{u} = du/dt$ where u is its displacement. Eventually we will be interested in distributed mass rather than point masses. Thus we can extend the definition to be

$$T = \frac{1}{2} \int_m v^2 dm = \frac{1}{2} \int_m \mathbf{v}^T \mathbf{v} \, dm$$

$$T = \frac{1}{2} \int_V \mathbf{v}^T \mathbf{v} \, \rho dV \tag{15.2}$$

where ρ is the mass per unit volume, $\mathbf{v}$ is the velocity vector, and V is the total volume containing the mass, m. Note that this form requires us to utilize the distribution of the velocity in space. Since the kinetic energy, T, is a scalar it is possible to split the volume into elements and compute

$$T = \sum_e T^e \tag{15.3}$$

where the kinetic energy of a typical element is

$$T^e = \frac{1}{2} \int_{V^e} (v^e)^2 \rho^e dV. \tag{15.4}$$

The speed distribution in space is related to the displacement distribution which we have previously treated. To illustrate this relation assume a one-dimensional continuum such that

$$v = \dot{u} = du/dt.$$

In a typical element we will assume a separation of variables such that

$$u(x,t) = H^e(x) \, \mathbf{u}^e(t) \tag{15.5}$$

This means that the unknown nodal displacements, $\mathbf{u}^e$, can now change with time, but we interpolate in space as before. This finite element interpolation assumption defines

$$T^e = \frac{1}{2} \int_{V^e} \dot{\mathbf{u}}^{e^T} \mathbf{H}^e(x)^T \mathbf{H}^e(x) \dot{\mathbf{u}}^e \rho^e \, dV.$$

But the $\dot{\mathbf{u}}^e$ can be taken outside the spatial integration to give a matrix form similar to Eq. (15.1):

$$T^e = \frac{1}{2} \dot{\mathbf{u}}^{e^T} \mathbf{m}^e \dot{\mathbf{u}}^e \qquad (15.6)$$

where the element **mass matrix** is given by

$$\mathbf{m}^e = \int_V \rho^e \mathbf{H}^{e^T} \mathbf{H}^e \, dV. \qquad (15.7)$$

It is a measure of the distribution of mass in space.

15.2.1 Point Masses

Previously we only considered a single type of element for a given problem class. Now we will have to consider two distinct types (structural elements and mass elements) during a single analysis. Usually the mass elements and structural elements will have common nodes. However, there are times where common engineering approximations treat them differently. For example, we sometimes refer to massless springs when we know that the mass of a spring is "small" compared to other masses in a system. Likewise another common approximation is that of a point mass.

In that case the mass is assumed to be lumped at a single point rather than being distributed in space. Then we do not need the above integral generalization for spatial masses. We could still use a matrix symbolism if desired. But the only non-zero entry in the mass matrix, $\mathbf{m}^e$, is the value of the point mass. Usually a point mass has no independent stiffness.

15.2.2 Mass Matrix Assembly

The mass matrix for the total system is obtained by the same direct assembly procedure that was used for the system stiffness matrix. The system mass and stiffness matrices have the same sparseness. However, it is not unusual for the mass distribution to be approximated as a diagonal matrix. In that case, the additional storage required by the system mass matrix is relatively small. Later we will examine how diagonal mass matrices can be produced.

15.3 Energy Method in Vibrations

For the undamped free vibration of a conservative system we can employ the principle of **conservation of energy.** Recall that since no energy is removed the total energy, E, must be constant. This consists of contributions from the kinetic energy, T, and the potential energy, π:

$$E = T + \pi = \text{Constant}. \tag{15.8}$$

Thus as T increases π decreases and visa versa. There are times when T is zero so that $E = \pi_{max}$. Likewise one can have $T_{max} = E$ when π is zero. This occurs when a system passes through its undeformed shape with its maximum speed. Since E is a constant we can write

$$\frac{d}{dt}(T + \pi) = 0. \tag{15.9}$$

For a general unforced system the potential energy is given by

$$\pi = \frac{1}{2}\mathbf{u}^T\mathbf{K}\,\mathbf{u} \tag{15.10}$$

and the kinetic energy is

$$T = \frac{1}{2}\dot{\mathbf{u}}^T\,\mathbf{M}\,\dot{\mathbf{u}} \tag{15.11}$$

where, as before, $\dot{\mathbf{u}} = d\mathbf{u}/dt$. Therefore,

$$\frac{d}{dt}\left(\frac{1}{2}\dot{\mathbf{u}}^T\,\mathbf{M}\,\dot{\mathbf{u}} + \frac{1}{2}\mathbf{u}^T\,\mathbf{K}\,\mathbf{u}\right) = 0. \tag{15.12}$$

In matrix notation this scalar equation reduces to

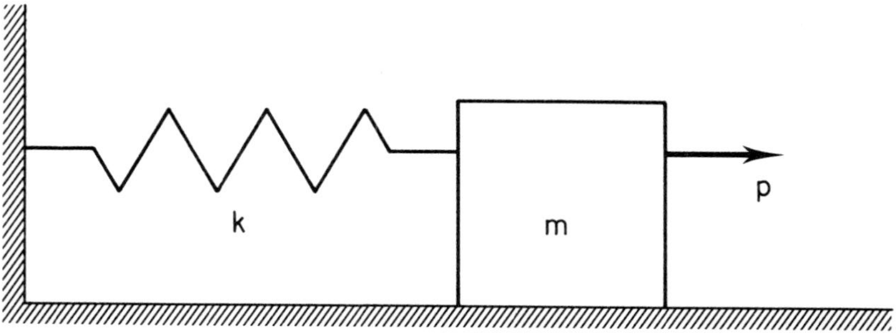

Fig. 15.1 A Spring-Mass System

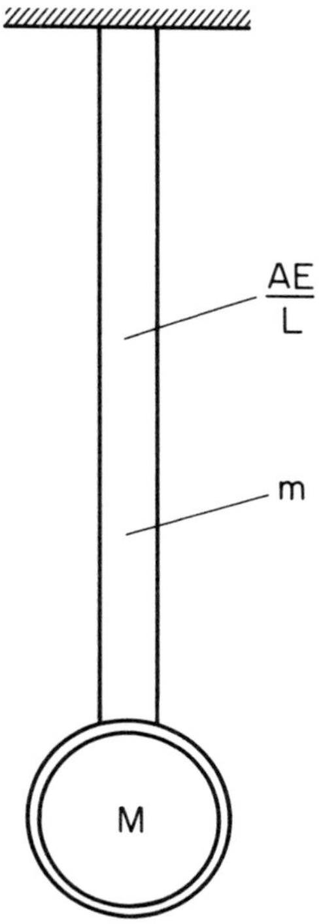

Figure 15.2 Elastic Rod with an End Mass

$$\frac{2}{2}\dot{\mathbf{u}}^T(\mathbf{M}\ \ddot{\mathbf{u}} + \mathbf{K}\ \mathbf{u}) = 0. \quad (15.13)$$

We will disregard the trivial solution that $\dot{\mathbf{u}} = \mathbf{0}$. This gives the equations system as

$$\mathbf{M}\ \ddot{\mathbf{u}} + \mathbf{K}\ \mathbf{u} = \mathbf{0}. \quad (15.14)$$

Since $\mathbf{K}\ \mathbf{u}$ can be thought of as the restoring force on a deformed system this represents the matrix form of **Newton's Second Law** of Motion. We will also want to consider the special form of this equation that is associated with simple harmonic motion. We assume that all of the displacement components, **u**, can be represented as

$$\mathbf{u}(t) = \mathbf{A}\sin(\omega t + p) \quad (15.15)$$

where **A** is an amplitude vector, t is time, p is a phase angle, and ω is the **frequency** of the oscillation. In this case the velocity and acceleration components are

$$\dot{\mathbf{u}}(t) = \omega\mathbf{A}\cos(\omega t + p) \quad (15.16)$$

and

$$\ddot{\mathbf{u}}(t) = -\omega^2\mathbf{A}\sin(\omega t + p). \quad (15.17)$$

Then Eq. (15.14) yields

$$(\mathbf{KA} - \omega^2\mathbf{MA})\sin(\omega t + p) = 0$$

or

$$(\mathbf{K} - \omega^2\mathbf{M})\mathbf{A} = \mathbf{0}. \quad (15.18)$$

If we assume that the mass matrix is invertable this could also be written as

$$(\mathbf{M}^{-1}\mathbf{K} - \omega^2\mathbf{I})\mathbf{A} = \mathbf{0}. \quad (15.19)$$

For a nontrivial solution, i.e. $\mathbf{A} \neq \mathbf{0}$, the determinant of the characteristic square matrix must vanish. For example,

$$|\mathbf{K} - \omega^2 \mathbf{M}| = 0. \tag{15.20}$$

For each eigenvalue, ω_i^2, that satisfies this relation there is a corresponding eigenvector, $\mathbf{A}_i$, of amplitudes.

15.3.1 Spring Point Mass System

The simplest dynamic system that we will consider is that of a massless spring and a single large point mass, M. The system structural contribution includes only the single spring. Thus the system stiffness matrix is

$$\mathbf{K} = k \begin{bmatrix} 1 & -1 \\ -1 & 1 \end{bmatrix}.$$

The point mass contributes only to the second node. Also, it generates a diagonal mass matrix. For this system the mass matrix is

$$\mathbf{M} = \begin{bmatrix} 0 & 0 \\ 0 & M \end{bmatrix}. \tag{15.21}$$

However, the first node is restrained and does not take part in the motion. Thus $u_1 \equiv 0$ so $A_1 \equiv 0$ and only u_2 (or A_2) remains active. Therefore, we need only consider the reduced partitions of $\mathbf{K}$ and $\mathbf{M}$ that are associated with u_2 (and A_2):

$$\mathbf{K}_r - \omega^2 \mathbf{M}_r = 0 \tag{15.22}$$

$$k - \omega^2 M = 0. \tag{15.23}$$

Solving this scalar equation gives the required frequency for the harmonic motion of this simple system:

$$\omega = \sqrt{k/M}\ . \tag{15.24}$$

This same result is usually derived from the differential equation of motion. The above quantity is called the **natural frequency.**

15.3.2 Elastic Rod with End Mass

A problem closely related to the previous one is to replace the spring with an elastic rod. As before, the axial stiffness from Fig. 15.2 is

$$\mathbf{K} = \mathbf{K}^e = \left(\frac{AE}{L}\right)^e \begin{bmatrix} 1 & -1 \\ -1 & 1 \end{bmatrix}.$$

But now we have to decide how to distribute the mass of the rod. Let the mass per unit length be ρ^e so its total mass is $m^e = \rho^e L^e$. Before when we considered uniformly distributed loads on a rod we assigned half of the resultant force to each end of the rod (see Eq. 4.65). We may be tempted to do the same with the uniformly distributed mass. If we create these two equal point masses we in effect assume a diagonal element mass matrix given by

$$\mathbf{m}^e{}_\ell = \frac{m^e}{2} \begin{bmatrix} 1 & 0 \\ 0 & 1 \end{bmatrix} \tag{15.25}$$

where the subscript ℓ denotes a point mass lumping. This type of mass matrix seems reasonable; but is it consistent with our previous assumptions for finite elements? To check this we recall the consistent integral definition:

$$\mathbf{m}^e = \int_L \mathbf{H}^{e^T} \mathbf{H}^e \, dm.$$

Here $dm = \rho dx$ and $\mathbf{H}^e$ for the rod was previously given in Eq. (4.18). The result of the integration is

$$\mathbf{m}^e = \frac{m^e}{6} \begin{bmatrix} 2 & 1 \\ 1 & 2 \end{bmatrix}. \tag{15.26}$$

Note that this consistent form is not diagonal, but it is symmetric with the same sparseness as $\mathbf{K}^e$.

Since a diagonal mass matrix has some desirable computational features some people recommend that the consistent form be somehow modified [10]. A simplistic approach is to simply add all row entries onto the diagonal position. That would convert Eq. (15.26) to Eq. (15.25). Another approach is to scale the diagonal of $\mathbf{m}^e$ such that the sum of the scaled diagonal terms is equal to the sum of all the original terms. That would also convert Eq. (15.26) to Eq. (15.25).

A mathematical approach might be to use the Lobatto rule (Table VI.II) and numerically integrate the consistent equation. That yields

$$\mathbf{m}^e = \rho L \left(\frac{1}{2}\mathbf{H}^{e^T}(x_1)\mathbf{H}^e(x_1) + \frac{1}{2}\mathbf{H}^{e^T}(x_2)\mathbf{H}^e(x_2)\right)$$

$$= \rho L\left(\begin{bmatrix} 1/2 & 0 \\ 0 & 0 \end{bmatrix} + \begin{bmatrix} 0 & 0 \\ 0 & 1/2 \end{bmatrix}\right) = \frac{m}{2}\begin{bmatrix} 1 & 0 \\ 0 & 1 \end{bmatrix}$$

as before.

For our first analysis we will consider the consistent form for the distributed mass. Assemblying the mass matrices gives

$$\mathbf{M} = \frac{m}{6}\begin{bmatrix} 2 & 1 \\ 1 & 2 \end{bmatrix} + M\begin{bmatrix} 0 & 0 \\ 0 & 1 \end{bmatrix} .$$

The kinetic energy of the system, from Eq. (15.11), is

$$T = \frac{1}{2}[\dot{u}_1 \ \ \dot{u}_2]\begin{bmatrix} m/3 & m/6 \\ m/6 & (m/3 + M) \end{bmatrix}\begin{Bmatrix} \dot{u}_1 \\ \dot{u}_2 \end{Bmatrix},$$

and the potential energy, from Eq. (15.10), is

$$\pi = \frac{1}{2}[u_1 \ \ u_2]\,\frac{AE}{L}\begin{bmatrix} 1 & -1 \\ -1 & 1 \end{bmatrix}\begin{Bmatrix} u_1 \\ u_2 \end{Bmatrix} .$$

But we have assumed that the first node is fixed. Thus $u_1 = 0$ and the above relations simplify to

$$T = 1/2(M + m/3)\,\dot{u}_2^2.$$

If we denote the reduced components of **u** as $\mathbf{u}_r$ this can be written as

$$T = \frac{1}{2}\dot{\mathbf{u}}_r^T \mathbf{M}_r \dot{\mathbf{u}}_r$$

$$\pi = \frac{1}{2}\mathbf{u}_r^T \mathbf{K}_r \mathbf{u}_r$$

where $\mathbf{M}_r = (M + m/3)$ and $\mathbf{K}_r = AE/L$. For **simple harmonic motion** the reduced form of Eq. (15.18) gives

$$\frac{AE}{L} - \omega^2(M + m/3) = 0$$

so that

$$\omega^2 = AE/(L(M + m/3))$$

is the approximation of the natural frequency (eigenvalue). If the rod were massless ($m = 0$) then this would be an exact result. If the end mass is not present ($M = 0$) our approximation is

$$\omega = \sqrt{3}\,\sqrt{\frac{AE}{mL}}$$

while the exact solution can be shown to be

$$\omega = \frac{\pi}{2}\sqrt{\frac{AE}{mL}} \quad .$$

So there would be about a 10% error in that case. When both M and m are present the exact value of ω is given by a transcendental equation. For $m = M$ our approximation would be within 1% of the exact result. When $m = 0.3$ M the error would be about 5%. Of course, as we increase the number of elements the magnitude of these errors will decrease.

Finally, if we had selected a lumped element mass matrix, the system mass matrix would change to

$$M = \begin{bmatrix} m/2 & 0 \\ 0 & (M + m/2) \end{bmatrix} \quad .$$

By inspection of the previous procedure one can write the approximation as

$$\omega^2 = AE/(L(M + m/2)).$$

15.4 Rayleigh's Method

The most common application of the Rayleigh method is to determine an approximate value for the lowest natural frequency, ω, of a conservative system. Of course, a system has the same number of natural frequencies as it has degrees of freedom. The problems illustrated above were approximated with only a single degree of freedom. To illustrate Rayleigh's method we again assume harmonic motion described by Eqs. (15.15) and (15.16). Substituting into Eq. (15.10) gives an alternate form of the potential energy:

$$\pi = \frac{1}{2}\mathbf{A}^T\mathbf{KA}\cos^2(\omega t + p)$$

$$= \pi^* \cos^2(\omega t + p)$$

while the corresponding kinetic energy is

$$T = \frac{1}{2}\mathbf{A}^T\mathbf{MA}\,\omega^2 \sin^2(\omega t + p)$$

$$= \omega^2 T^* \sin^2(\omega t + p).$$

Returning to Eq. (15.8), and substituting $\sin^2 = 1 - \cos^2$ gives

$$E = \pi + T = \text{const.}$$

$$= (\pi^* - \omega^2 T^*)\cos^2(\omega t + p) + \omega^2 T^*.$$

The terms π^* and T^* depend only on the relative spatial amplitudes (mode shapes), **A**, and do not vary with time. For the time derivative to vanish, as in Eq. (15.9), we find that

$$\pi^* - \omega^2 T^* = 0$$

thus

$$\omega^2 = \frac{\pi^*}{T^*} = \frac{\frac{1}{2}\mathbf{A}^T\mathbf{KA}}{\frac{1}{2}\mathbf{A}^T\mathbf{MA}}. \qquad (15.27)$$

This is known as the **Rayleigh quotient.**

If we know the exact i-th mode shape or eigenvector, $\mathbf{A}_i$, of a harmonic motion the Rayleigh quotient will give the corresponding frequency, ω_i. It can be shown that the lowest value of the Rayleigh quotient corresponds to the first mode, or "the" natural frequency. Any errors in the mode shape, **A**, cause ω to be overestimated.

The deformed static equilibrium shape is commonly used as a reasonable approximation of the first mode. Thus some static analysis programs also give an (upper bound) estimate of the natural frequency of the system.

15.4.1 Spring-Mass System

To illustrate the use of the Rayleigh quotient we will return to the simple spring-mass system. For a single (active) degree of freedom system; that is, for a single element system we can easily estimate the mode shape. Since we assumed that the first node was fixed we let $\mathbf{A}^T = [0 \quad A_2]$ so

$$\pi^* = \frac{1}{2}\mathbf{A}^T\mathbf{K}\mathbf{A}$$

$$= \frac{1}{2}[0 \quad A_2]\begin{bmatrix} k & -k \\ -k & k \end{bmatrix}\begin{Bmatrix} 0 \\ A_2 \end{Bmatrix}$$

$$= \frac{1}{2}kA_2^2$$

and

$$T^* = \frac{1}{2}\mathbf{A}^T\mathbf{M}\mathbf{A}$$

$$= \frac{1}{2}[0 \quad A_2]\begin{bmatrix} 0 & 0 \\ 0 & M \end{bmatrix}\begin{Bmatrix} 0 \\ A_2 \end{Bmatrix}$$

$$= \frac{1}{2}MA_2^2.$$

Thus the Rayleigh quotient for the corresponding (natural) frequency is

$$\omega^2 = \pi^*/T^* = k/M$$

which agrees with the previous result in Eq. (15.22).

As a final example of how static deflections can be used to estimate the natural frequency reconsider the elastic rod without the large end mass. We would like to introduce more elements and more degrees of freedom in an attempt to improve our natural frequency estimate given previously.

Select two equal rod elements, each with a length of $\ell = L/2$. The degrees of freedom for the system are $\mathbf{u}^T = [u_1 \quad u_2 \quad u_3]$ (or $\mathbf{A}^T = [A_1 \quad A_2 \quad A_3]$). The assembled system stiffness matrix is

$$\mathbf{K} = \frac{AE}{\ell}\begin{bmatrix} 1 & -1 & 0 \\ -1 & 1+1 & -1 \\ 0 & -1 & 1 \end{bmatrix}$$

and the consistent system mass matrix is

$$M = \frac{\rho A \ell}{6}\begin{bmatrix} 2 & 1 & 0 \\ 1 & 2+2 & 1 \\ 0 & 1 & 2 \end{bmatrix} .$$

To estimate ω we will use a static deflection shape for the rod. We have fixed node 1 so $u_1 = 0 = A_1$. Clearly the maximum deflection will occur at the free end (u_3). Since we do not know the eigenvector, **A**, we will estimate it. If we use the static deflected shape for a concentrated end load we would have $u_2 = u_3/2$. But, that load condition does not consider the gravity (mass) effects. If we consider the static deflection for a rod loaded by its own weight we know from mechanics of materials that the static deflection is

$$u(x) = u_{max}[2(x/L) - (x/L)^2].$$

Thus the more reasonable estimates for the relative deformations are $u_2 = 3u_{max}/4$ and $u_3 = u_{max}$. Thus we will assume the same ratios for the eigenvector, **A**. Let $\mathbf{A}^T = [0 \;\; 3 \;\; 4]$. Then we have

$$\pi^* = \frac{1}{2}[0 \;\; 3 \;\; 4]\,\frac{AE}{\ell}\begin{bmatrix} 1 & -1 & 0 \\ -1 & 2 & -1 \\ 0 & -1 & 1 \end{bmatrix}\begin{Bmatrix} 0 \\ 3 \\ 4 \end{Bmatrix}$$

$$\pi^* = \frac{10AE}{2\ell} = \frac{10AE}{L}$$

and

$$T^* = \frac{1}{2}[0 \;\; 3 \;\; 4]\,\frac{\rho A \ell}{6}\begin{bmatrix} 2 & 1 & 0 \\ 1 & 4 & 1 \\ 0 & 1 & 2 \end{bmatrix}\begin{Bmatrix} 0 \\ 3 \\ 4 \end{Bmatrix}$$

$$T^* = \frac{46\rho A \ell}{6} = \frac{23\rho AL}{6}.$$

Noting that $m = \rho AL$ our approximation becomes

$$\omega = \sqrt{\frac{60}{23}} \sqrt{\frac{AE}{mL}} .$$

The correct constant is $\pi/2 = 1.571$. Our single element estimate was 1.732 and the above value for the static shape is 1.615. Clearly the accuracy is increasing with the number of degrees of freedom. If we had selected the diagonal mass matrix the two element approximation would give a constant of 1.534.

15.5 Exercises

1. Use the Rayleigh quotient method to estimate the natural frequency of an axial bar. Utilize three equally spaced elements.

References and bibliography

1. Akin, J. E., **Application and Implementation of Finite Element Methods**, Academic Press, London, 1982.
2. Akin, J. E. and Gray, W. H., Contouring on Isoparametric Surfaces, **Int. J. Num. Meth. Eng.**, Dec. 1977.
3. Akin, J. E. and Wooten, J. W., Tokamak Plasma Equilibria by Finite Elements, R. H. Gallagher (Ed.), **Finite Elements in Fluids III**, John Wiley, New York, Chapter 21, 1978.
4. Akin, J. E., Verification Checks of Finite Element Models, **Computing in Civil Engineering**, A.S.C.E., 43-50, New York, NY 1981.
5. Barlow, J., Optimal Stress Locations in Finite Element Models, **Int. J. Num. Meth. Eng.**, 10, 243-51, 1976.
6. Becker, E. B., Carey, G. F. and Oden, J. T., **Finite Elements An Introduction**, Prentice Hall, Englewood Cliffs, 1981.
7. Chung, T. J., **Finite Element Analysis in Fluid Dynamics**, McGraw-Hill, New York, 1978.
8. Connor, J. C. and Brebbia, C. A., **Finite Element Techniques for Fluid Flow**, Butterworth, 1976.
9. Cook, R. D., **Concepts and Applications of Finite Element Analysis**, John Wiley, New York, 1974.
10. Desai, C. S., **Elementary Finite Element Method**, Prentice Hall, Englewood Cliffs, 1979.
11. Fenner, R. T., **Finite Element Methods for Engineers**, Macmillan Ltd., London, 1975.
12. Gallagher, R. H., **Finite Element Analysis Fundamentals**, Prentice-Hall, Englewood Cliffs, 1975.
13. Hinton, E. and Owen, D. R. J., **Finite Element Programming**, Academic Press, London, 1977.
14. Huebner, K. H., **Finite Element Method for Engineers**, John Wiley, New York, 1974.
15. Martin, H. C. and Carey, G. F., **Introduction to Finite Element Analysis**, McGraw-Hill, New York, 1974.
16. Myers, G. E., **Analytical Methods in Conduction Heat Transfer**, McGraw-Hill, New York, 1971.
17. Norrie, D. H. and DeVries, G., **Finite Element Method: Fundamentals and Applications**, Academic Press, New York, 1973.
18. Norrie, D. H. and DeVries, G., **Finite Element Bibliography**, Plenum Press, New York, 1976.
19. Rockey, K. C., et al., **Finite Element Method - A Basic Introduction**, Halsted Press, 1975.

20. Segerlind, L. J., **Applied Finite Element Analysis**, John Wiley, New York, 1984.
21. Ural, O., **Finite Element Method: Basic Concepts and Application**, Intext, 1973.
22. Weaver, W. F. Jr., and Johnston, P. R., **Finite Elements for Structural Analysis**, Prentice Hall, Englewood Cliffs, 1984.
23. Whiteman, J. R., **A Bibliography of Finite Elements**, Academic Press, London, 1975.
24. Zienkiewicz, O. C., **The Finite Element Method**, McGraw-Hill, London, 1977.

Index

A

B

C

D

E

F

G

H

I

J

K

L

M

$$\mathbf{C} = \frac{QAt}{3}\begin{Bmatrix} 1 \\ 1+1+1 \\ 1+1+1 \\ 1 \\ 1+1+1 \\ 1 \end{Bmatrix} + \begin{Bmatrix} 0 \\ 0 \\ 0 \\ q_4 \\ q_5 \\ q_6 \end{Bmatrix} = \begin{Bmatrix} 4 \\ 12 \\ 12 \\ 4 \\ 12 \\ 4 \end{Bmatrix} + \begin{Bmatrix} 0 \\ 0 \\ 0 \\ q_4 \\ q_5 \\ q_6 \end{Bmatrix} .$$

In the above vector the q's are the nodal heat flux reactions required to maintain the specified external temperature. Since the last three equations have essential boundary conditions applied we can reduce the first three to

$$\begin{bmatrix} 4 & -4 & 0 \\ -4 & 16 & -8 \\ 0 & -8 & 16 \end{bmatrix}\begin{Bmatrix} T_1 \\ T_2 \\ T_3 \end{Bmatrix} = \begin{Bmatrix} 4 \\ 12 \\ 12 \end{Bmatrix}$$

$$-T_4\begin{Bmatrix} 0 \\ -4 \\ 0 \end{Bmatrix} - T_5\begin{Bmatrix} 0 \\ 0 \\ -8 \end{Bmatrix} - T_6\begin{Bmatrix} 0 \\ 0 \\ 0 \end{Bmatrix} . \tag{12.20}$$

Substituting the data that the exterior surface temperature is five yields the reduced source term

$$\mathbf{C^*} = \begin{Bmatrix} 4 \\ 12 \\ 12 \end{Bmatrix} + \begin{Bmatrix} 0 \\ 20 \\ 0 \end{Bmatrix} + \begin{Bmatrix} 0 \\ 0 \\ 40 \end{Bmatrix} = \begin{Bmatrix} 4 \\ 32 \\ 52 \end{Bmatrix} .$$

Solving for the interior temperatures, Fig. 2.3 gives the inverse

$$\mathbf{S^*}^{-1} = \frac{1}{512}\begin{bmatrix} 192 & 64 & 32 \\ 64 & 64 & 32 \\ 32 & 32 & 48 \end{bmatrix} \tag{12.21}$$

so multiplying by $\mathbf{C^*}$ yields:

$$\mathbf{T^*} = \begin{Bmatrix} 8.750 \\ 7.750 \\ 7.125 \end{Bmatrix} . \tag{12.22}$$

Substituting these values into the original system equations will give the exterior nodal heat flux values (**thermal reactions**) required by this problem. For example, the fourth equation yields:

N

O

P

Q

R

S

T